NATIONAL ACADEMIES
Sciences
Engineering
Medicine

Engaging Scientists in Central Asia on Life Science Data Governance Principles

Anne Johnson and Trisha Tucholski, *Rapporteurs*

Board on Life Sciences

Division on Earth and Life Studies

International Networks and Cooperation

Policy and Global Affairs

Proceedings of a Workshop Series

NATIONAL ACADEMIES PRESS 500 Fifth Street, NW Washington, DC 20001

This activity was supported by AWD-001082 between the National Academy of Sciences and the U.S. State Department. Any opinions, findings, conclusions, or recommendations expressed in this publication do not necessarily reflect the views of any organization or agency that provided support for the project.

International Standard Book Number-13: 978-0-309-70662-9
International Standard Book Number-10: 0-309-70662-9
Digital Object Identifier: https://doi.org/10.17226/27156

This publication is available from the National Academies Press, 500 Fifth Street, NW, Keck 360, Washington, DC 20001; (800) 624-6242 or (202) 334-3313; http://www.nap.edu.

Printed in the United States of America.

Suggested citation: National Academies of Sciences, Engineering, and Medicine. 2024. *Engaging Scientists in Central Asia on Life Science Data Governance Principles: Proceedings of a Workshop Series*. Washington, DC: The National Academies Press. https://doi.org/10.17226/27156.

The **National Academy of Sciences** was established in 1863 by an Act of Congress, signed by President Lincoln, as a private, nongovernmental institution to advise the nation on issues related to science and technology. Members are elected by their peers for outstanding contributions to research. Dr. Marcia McNutt is president.

The **National Academy of Engineering** was established in 1964 under the charter of the National Academy of Sciences to bring the practices of engineering to advising the nation. Members are elected by their peers for extraordinary contributions to engineering. Dr. John L. Anderson is president.

The **National Academy of Medicine** (formerly the Institute of Medicine) was established in 1970 under the charter of the National Academy of Sciences to advise the nation on medical and health issues. Members are elected by their peers for distinguished contributions to medicine and health. Dr. Victor J. Dzau is president.

The three Academies work together as the **National Academies of Sciences, Engineering, and Medicine** to provide independent, objective analysis and advice to the nation and conduct other activities to solve complex problems and inform public policy decisions. The National Academies also encourage education and research, recognize outstanding contributions to knowledge, and increase public understanding in matters of science, engineering, and medicine.

Learn more about the National Academies of Sciences, Engineering, and Medicine at **www.nationalacademies.org**.

Consensus Study Reports published by the National Academies of Sciences, Engineering, and Medicine document the evidence-based consensus on the study's statement of task by an authoring committee of experts. Reports typically include findings, conclusions, and recommendations based on information gathered by the committee and the committee's deliberations. Each report has been subjected to a rigorous and independent peer-review process and it represents the position of the National Academies on the statement of task.

Proceedings published by the National Academies of Sciences, Engineering, and Medicine chronicle the presentations and discussions at a workshop, symposium, or other event convened by the National Academies. The statements and opinions contained in proceedings are those of the participants and are not endorsed by other participants, the planning committee, or the National Academies.

Rapid Expert Consultations published by the National Academies of Sciences, Engineering, and Medicine are authored by subject-matter experts on narrowly focused topics that can be supported by a body of evidence. The discussions contained in rapid expert consultations are considered those of the authors and do not contain policy recommendations. Rapid expert consultations are reviewed by the institution before release.

For information about other products and activities of the National Academies, please visit www.nationalacademies.org/about/whatwedo.

ENGAGING SCIENTISTS IN CENTRAL ASIA ON LIFE SCIENCE DATA GOVERNANCE

WORKSHOP SERIES PLANNING COMMITTEE

LUSINE POGHOSYAN (*Chair*), Columbia University, United States
DAMIRA OMURZAKOVNA ASHIRALIEVA, National Scientific-Practical Center, Ministry of Health, Kyrgyzstan
SHOLPAN ASKAROVA, Nazarbayev University, Kazakhstan
MARGARITA YULAYEVNA ISHMURATOVA, Buketov Karaganda University, Kazakhstan
YANN JOLY, McGill University, Canada
FAINA LINKOV, Duquesne University, United States
JAILOBEK OROZOV, A. Duisheev Kyrgyz Research Institute of Veterinary Science, Kyrgyzstan
VASILIKI RAHIMZADEH, Baylor College of Medicine, United States
KRYSTAL TSOSIE, Arizona State University, United States
JOHN COLIN URE, Access Partnership, Singapore
GAUTHAM VENUGOPALAN, Gryphon Scientific, United States
WEI ZHENG, Vanderbilt University, United States

Staff

TRISHA TUCHOLSKI, Program Officer, Board on Life Sciences
KAVITA BERGER, Director, Board on Life Sciences
RITA GUENTHER, Senior Program Officer, International Networks and Cooperation
CARMEN SHAW, Associate Program Officer, International Networks and Cooperation
NAM VU, Senior Program Assistant, Board on Life Sciences

Reviewers

This proceedings of a workshop series was reviewed in draft form by individuals chosen for their diverse perspectives and technical expertise. The purpose of this independent review is to provide candid and critical comments that will assist the National Academies of Sciences, Engineering, and Medicine in making each published proceedings as sound as possible and to ensure that it meets the institutional standards for quality, objectivity, evidence, and responsiveness to the charge. The review comments and draft manuscript remain confidential to protect the integrity of the process.

We thank the following individuals for their review of this proceedings:

Yann Joly, McGill University, Canada
Margarita Yulayevna Ishmuratova, Buketov Karaganda University, Kazakhstan
Namazbek Kudaibergenovich Abdykerimov, National Academy of Sciences of Kyrgyzstan
Kalysbek Kydyshov, Ministry of Health of Kyrgyzstan
Nurbolot Usenbaev, Ministry of Health of Kyrgyzstan
Zulfiya Burievna Davlyatnazarova, National Academy of Sciences of Tajikistan
Anvar Kodirov, National Academy of Sciences of Tajikistan
Tatyana Novossiolova, Center for the Study of Democracy, Bulgaria

Although the reviewers listed above provided many constructive comments and suggestions, they were not asked to endorse the content of the proceedings nor did they see the final draft before its release. The review of this proceedings was overseen by coordinator **Margaret E. Kosal** of Georgia Institute of Technology. She was responsible for making certain that an independent examination of this proceedings was carried out in accordance with standards of the National Academies and that all review comments were carefully considered. Responsibility for the final content rests entirely with the rapporteurs and the National Academies.

Acknowledgments

The workshop planning committee and the National Academies of Sciences, Engineering, and Medicine staff team would like to thank our colleagues at the State Department and the International Science and Technology Center in Central Asia for their assistance in soliciting experts and sharing information about the project. We would also like to thank the National Academies Research Center for early assistance in a search for subject matter experts for this project. Finally, we would like to thank our interpreters, translators, and editors from Language Exchange Translations, LLC, specifically Irina Krasnokutsky, Irina Paramonova, and Pavel Palazhchenko and their colleagues, whose exceptional expertise and professionalism facilitated communications across languages and cultures, and ensured that this project was truly collaborative from start to finish.

Contents

FIGURES

Acronyms and Abbreviations

DEPA	Data Empowerment and Protection Architecture
EU	European Union
FAIR	Findability, Accessibility, Interoperability, Reusability
GA4GH	Global Alliance for Genomics and Health
GBIF	Global Biodiversity Information Facility
GDPR	General Data Protection Regulation
iGEM	International Genetically Engineered Machine
ISTC	International Science and Technology Center
NIH	U.S. National Institutes of Health
PHA4GE	Public Health Alliance for Genomic Epidemiology
UNESCO	United Nations Educational, Scientific and Cultural Organization
WHO	World Health Organization

Overview

The rapid proliferation of life science data presents tremendous opportunities to address long-standing and emerging challenges in health, agriculture, the environment, and other sectors. Data from scientific research represent the outcomes of significant investment on the part of individuals and communities, institutions, industries, and governments. Some data generated and collected are sensitive, often warranting special protections. Life science data can include, for example, personal and genetic data, proprietary data, and/or data that may be otherwise harmful if accidentally or intentionally misused. Different types of life science data (e.g., genetic vs. nongenetic, human vs. animal) may necessitate different considerations to appropriately balance risks and benefits of data sharing and protection. To facilitate scientific progress and collaboration while also protecting personal privacy, the integrity of research investments, and national security, it is important for life science data to be collected, digitized, secured, and shared responsibly.

To exchange experiences and collaboratively discuss best practices for protecting life science data and promoting responsible data sharing in Central Asia, the U.S. National Academies of Sciences, Engineering, and Medicine convened a workshop series titled "Engaging Scientists in Central Asia on Life Science Data Governance Principles," held virtually across six 2-hour sessions in May and June 2023. The workshop series brought together early-career and established scientists, along with experts in data science, cybersecurity, and law, from Kazakhstan, Kyrgyzstan, Tajikistan, the United States, Uzbekistan, and elsewhere to explore existing policies and practices, gaps and limitations, fundamental concepts relevant to data governance, and opportunities for improvement in data sharing practices and governance structures. Throughout the workshop series, participants considered how concepts and approaches relevant to data governance fit into the overall research context to help minimize the risks and maximize the benefits of life science research endeavors in Central Asia and elsewhere. In addition, the National Academies commissioned seven papers by authors from Kazakhstan, Kyrgyzstan, and Tajikistan on a variety of issues related to data governance. These papers are provided in Appendix E of this report.

Data governance has been defined as "the principles, procedures, frameworks, and policies that ensure acceptable and responsible processing of data at each stage of the data life cycle... [to] maintain data integrity, quality, availability, accessibility, usability, and security" (Eke et al., 2022, p. 600-601). Data governance is important not only because researchers have a responsibility to protect personal privacy and the integrity of scientific investments, but also because good data governance makes science more impactful. Each stage of the data life cycle—including data collection, storage, processing, curation, sharing, application and use, and deletion—has unique considerations that impact data governance. For instance, during data *collection*, researchers consider such issues as whether informed consent for sharing and using the data has been obtained, whether biases exist for sampling, and whether there are legal and/or regulatory differences that affect sharing the data between jurisdictions. During data *storage*, considerations include how to safely and securely collect and handle data, how long data should be preserved, who owns the rights to the data, who pays for data storage, and how to minimize risks of data leaks as technology changes over time. In the *application and use* of data, researchers consider how to communicate incidental findings that pertain to the health of study participants, how to ensure that the minimal amount of data is used to reduce risk to participants, how to ensure data are not misused, and how to ensure data are used responsibly for civil and military applications or for economic gain (Eke et al., 2022). "Good governance frameworks really achieve two aims: They maximize the scientific utility of that data, as well as the social value of the research that's derived from that data," said committee member Vasiliki Rahimzadeh, Baylor College of Medicine (United States). She noted that responsible governance frameworks help governments and institutions use the outcomes of scientific research to make informed decisions for the benefit of all members of society.

By providing a forum for open exchange of information, experiences, and views on data sharing and protection, the workshop series helped elucidate several challenges and opportunities facing researchers who seek to maximize the value of their research investments for scientific progress while upholding values such as respect, fairness, safety, and security. "Now more than ever, life scientists are collecting and analyzing vast amounts of data from various sources," said Lusine Poghosyan, Columbia University (United States), chair of the workshop's planning committee. "Digitalization of these data presents new opportunities to address significant challenges across various sectors and accelerate research through collaboration within and across borders. It's important to ensure that data are shared responsibly, and that appropriate safeguards are in place to protect the privacy of individuals and the investments made by institutions, industries, governments, and others."

In sessions highlighting particular scientific topics being pursued in Central Asia, participants explored both the benefits of scientific advances and the risks and vulnerabilities involved in collecting and sharing biological data and knowledge. In the first workshop, participants presented a sampling of life science research efforts in Central Asia. Scientists from Kazakhstan, Kyrgyzstan, and Uzbekistan, working in veterinary medicine, infectious disease, antimicrobial resistance, genomics, botany, and other areas, shared their successes and challenges in conducting life science research, digitizing data, and contributing to international databases. They pointed to some key challenges facing the region's research efforts, with common limitations including a dearth of comparison data from the region, difficulty accessing biobanks and other technological resources, and limited expertise in some areas such as data science.

While there are currently no international data governance laws, several organizations offer resources and guidelines for responsible data sharing across borders. These include the United Nations Educational, Scientific and Cultural Organization; the World Health Organization; and the Global Alliance for Genomics and Health. At the second workshop, several participants suggested that successful international collaborations hinge on developing and adopting data interoperability standards and harmonizing national data sharing and security laws. Participants examined existing international and national policies, practices, and norms related to access and transfer of biological and life science data, focusing on data sharing and protection. Data governance experts from Germany, India, Taiwan, Uganda, and the United States described their data sharing laws and norms as potential models for researchers in other countries to learn from when developing, evaluating, and improving upon their own frameworks. Some participants also attested that their current national data governance standards are in flux. During these discussions, speakers also described challenges in data management and international data sharing, such as the cost of long-term data storage and the lack of advanced infrastructure and subject matter expertise in some regions.

The goal of the third workshop was to provide a balanced view of biological data sharing. Globally, there is a shift toward a culture of open science, which is widely viewed as the key to advancing life science research. However, in practice, hurdles—such as real and perceived risks and harms for participants—prevent biological data from being shared freely and openly. In the third workshop, the benefits of sharing biological data were considered together with the risks and vulnerabilities of doing so, in order to explore how data can be shared more feasibly, equitably, and securely. As an example, several speakers noted that tribal and Indigenous communities worldwide can be harmed by unethical research practices; in response, Indigenous communities are exploring methods of protecting and stewarding the data and knowledge their communities hold, in order to have greater input into decisions on how data from their community members are collected, stored, and shared, and to have a greater share of the benefits derived from research using those data and knowledge, especially when the benefits directly affect their communities. Each of the five Central Asian countries has communities with unique traditional knowledge, which not only holds considerable personal and spiritual meaning but also may be important for the survival of these communities. Several speakers suggested that stronger legal protections may be needed to preserve traditional knowledge and noted that access to resources and legal expertise could help address inequities in the ability for communities to navigate intellectual property and patent law. Participants discussed different risk, use, and sharing considerations that apply to a variety of data types. Different perspectives on the definitions of and roles for ethics, equity, and equality in science were also shared.

Participants also explored life science data governance principles in practice. In the fourth workshop, speakers shared experiences with the existing national practices, policies, and norms around protecting and sharing life science data in Central Asia; described unique features, common challenges, and common goals; and explored opportunities to improve data governance policies and implementation. Existing practices address electronic data governance, conflicts of interest, personal data, biometric registration, genetic resources, and centralized digitization efforts. These policies and norms are often applied in a project-centric way, and researchers are typically required to secure their data and publish or deposit them into international repositories. Emphasizing the importance of standardized protocols and appropriate security measures to support data sharing, panelists from Kyrgyzstan, Tajikistan, and Uzbekistan identified common challenges with data sharing in their countries, including limitations in available data and computing power; a lack of expertise in cybersecurity, artificial intelligence, and digitization; and a need for improved digital infrastructure.

The workshops also surfaced opportunities and challenges in terms of the ability to translate data governance practices in different institutional contexts. The fifth workshop focused on managing cybersecurity risks with the goal of maintaining the confidentiality, integrity, and availability of life science data. Speakers commented that a commitment to open science includes a responsibility to share data in a secure and meaningful way to advance science, but the methods employed to achieve this vary depending on an institution's context, goals, and resources. Cyberattacks can disrupt research and compromise sensitive data, and it is therefore important to balance open data sharing with risk assessment, especially when international collaborations or dual-use research are involved. Speakers discussed several approaches to advance cybersecurity in scientific organizations, including basic cyber hygiene, resources, and practices, along with more intensive cyber resilience strategies. Panelists also emphasized the importance of vigilance in the face of social engineering methods, such as phishing, that can manipulate staff into divulging sensitive information, such as passwords and personal information.

In the sixth and final workshop, organizers summarized what was learned from the previous five workshops and facilitated a collaborative discussion to explore areas on which to build. Participants highlighted concrete suggestions that decision-makers in Central Asian countries can implement to improve their data governance frameworks, infrastructure, and expertise, thereby allowing researchers to participate in international collaborations more fully. One recurring theme throughout the workshop series was the idea that researchers have a duty to share data responsibly in an effort to advance science and that this means recognizing cybersecurity threats and implementing appropriate access controls and data protections. Working with, protecting, and sharing data require time and resources, however. As Yann Joly (committee member), McGill University (Canada), stated, "the devil's in the detail[s]." Joly reiterated that creating an agreed-upon international data governance framework requires work: integrating and harmonizing the disparate national laws; creating incentives to motivate scientists to follow responsible data sharing practices; and designing clear policies that protect subjects' and researchers' data, intellectual property, and national security.

Looking forward, workshop series planning committee members highlighted four ways participants and other experts can implement and advance the life science data governance best practices that were described during the workshop series. First, participants can create and support regional or national scientific networks consisting of data repositories, scientific societies, journals, and conferences, some of which already exist or are being built. Second, funding sources can be sought for in-person meetings and sustained engagement, initiatives, and collaborations. Third, Central Asian scientists can be encouraged to publish in journals and use repositories that are available internationally, with help from international researchers who can offer scientific communication training to overcome language or cultural barriers. Finally, engaging diverse groups of scientists, students, decision-makers, ministry officials, legal experts, and other relevant groups in creating a more supportive environment for scientific collaboration can help to expand the reach and benefits of research. Participants closed the final workshop session by discussing follow-on activities being planned to disseminate the information and views shared during the workshop series along with future opportunities to continue the conversation with a broader set of participants.

Introduction

The rapid proliferation of life science data presents tremendous opportunities to address long-standing challenges in health, agriculture, the environment, and other sectors. However, many of these data are sensitive and represent significant investment on the part of individuals and communities, institutions, industries, and governments. Furthermore, different types of life science data (e.g., genetic vs. nongenetic, human vs. animal) raise different considerations for data sharing and protection. To maximize the benefits of data sharing, protect the integrity of research investments, and minimize risks to personal privacy and national security, it is important for data to be collected, digitized, secured, and shared responsibly.

To exchange experiences and collaboratively explore best practices for protecting life science data and promoting responsible data sharing in Central Asia, the U.S. National Academies of Sciences, Engineering, and Medicine convened a workshop series titled "Engaging Scientists in Central Asia on Life Science Data Governance Principles," held virtually across six 2-hour sessions in May and June 2023. The workshops convened early-career and established scientists, legal experts, and data science and cybersecurity experts from Central Asia, the United States, and elsewhere to explore current practices and limitations, goals and concepts relevant to advancing data governance, and opportunities for improvement.

The statement of task for this activity can be found in Appendix A; agendas for each workshop are provided in Appendix B; and Appendix C provides biographies of the workshop planning committee members. Appendix D presents a list of individuals who participated in one or more sessions, organized by country. Seven papers commissioned for this workshop are available in Appendix E. And a list of international and national resources shared throughout the session can be found in Appendix F. Each consecutive workshop built on the previous workshop to ensure continuity of discussion and maximize the benefit to participants. The first workshop set the stage with a sampling of life science research efforts in the Central Asian countries of Kazakhstan, Kyrgyzstan, Uzbekistan, and Tajikistan. The second workshop examined data governance principles and practices in different countries and explored their role in international research collaborations. In the third workshop, participants considered the benefits and risks of sharing biological data, and in the fourth workshop, participants reviewed existing national practices, policies, and norms around life science data sharing in Kazakhstan, Kyrgyzstan, and Tajikistan. The fifth workshop focused on specific practices relevant to managing cybersecurity risks. In the sixth and final workshop, participants reviewed key takeaways from the previous sessions, described examples of key needs and opportunities for enhancing data governance for life science in Central Asia, and discussed future opportunities to further examine these issues and inform decision-making to ultimately improve existing data governance policies and practices.

This proceedings of a workshop series summarizes the discussions at each of these six workshops. It was produced by rapporteurs on behalf of the National Academies based on recordings, slides, and transcripts from the workshops. Participants were informed that by participating in the workshop series, they consented to their comments being recorded and used for the purposes of this proceedings. It is intended as a factual reflection of the discussions at the workshops and does not represent consensus views or recommendations of the National Academies. Comments or viewpoints reflected in this proceedings should be attributed to the individual, not their organization, unless otherwise stated.

1
The Life Science Research Landscape in Central Asia

To launch the workshop series, Trisha Tucholski, U.S. National Academies of Sciences, Engineering, and Medicine, and Lusine Poghosyan, Columbia University (United States), welcomed participants and outlined the goals of the first workshop, held on May 4, 2023. Setting the stage for the discussions to come, the speakers explained that this workshop series was designed to highlight life science and biological research efforts across Kazakhstan, Kyrgyzstan, Uzbekistan, and Tajikistan, explore the scale and scope of life science and biological data in the region, understand national and institutional data practices and norms, discuss existing and imminent challenges related to life science data governance, and explore emerging data governance tools and trends.

LIFE SCIENCE DATA GOVERNANCE: PRINCIPLES, POLICIES, AND PRACTICES

Vasiliki Rahimzadeh, Baylor College of Medicine (United States), offered an introduction to general principles, policies, and practices relevant to life science data governance in Central Asia. Defining *life science data* as "any data that are related to living organisms and their biological processes," she underscored the value of these data for advancing understanding of fundamental biology, improving human health, and preserving the Earth's natural resources. This value depends upon the proper and ethical collection, management, and analysis of data; to this end, she said, those data governance frameworks are needed to advance scientific research, protect and share data, and reflect the values and priorities of communities in Central Asia and elsewhere.

Three central concepts that underpin frameworks for responsible sharing of life science data include privacy, protection, and governance. *Privacy* is the right to control access to information about oneself. The right to privacy is activated through *protections*, which generally refer to the collection of laws, policies, and procedures that are applied to data to minimize unauthorized intrusion. Data privacy and protections are the foundation of good *governance*, which means applying rules to ensure authorized access to stored data, security (standards and technologies to secure data), management (collective storing and sharing processes), and stewardship (functional practices ensuring that data assets are appropriately protected and used according to the values and wishes of those who contributed them).

Rahimzadeh highlighted several attributes of data governance frameworks. First, responsible data governance frameworks respect people. Health, genetic, and genomic data are rich resources for learning more about health behavior, status, and potential preventive practices, but the collection and dissemination of such data also can confer serious risks for individuals, including the risk of reidentification even after biological data have been separated from personally identifiable information. Many national data protection laws address these risks, but the differences among such laws can complicate global life science research. Responsible governance frameworks also advance ethical research practices.

Rahimzadeh noted that maximizing the scientific utility of life science data—often generated in extremely large quantities—requires appropriate storage capacity and computing power, but also a recognition that the data points represent real people who deserve respect and expect their data to be put to good use. "Good governance frameworks really achieve two aims: They maximize the scientific utility of those data, as well as the social value of the research that's derived from those data," she said. Finally, she noted that responsible governance frameworks help governments and institutions make informed decisions about how to best preserve the Earth's resources and habitats for all life.

LIFE SCIENCE RESEARCH IN CENTRAL ASIA

Damira Ashiralieva, National Scientific-Practical Center, Ministry of Health of Kyrgyzstan, moderated a panel highlighting research efforts in Kazakhstan, Kyrgyzstan, Uzbekistan, and Tajikistan on topics ranging from biomedicine to animal diseases to biodiversity. Together, these presentations provided a window into the scale and scope of life science and biological research efforts in the region, along with perspectives on data governance practices and challenges.

Challenges and Best Practices for Engaging Scientists in Central Asia

Faina Linkov, Duquesne University (United States), provided a brief overview of the public health context in which biomedical research is carried out in Central Asia, along with some key challenges related to improving the region's research infrastructure and data governance practices.

Countries in Central Asia face high rates of both chronic and infectious diseases, such as lung cancer, brucellosis, and cirrhosis, which contribute to a lower life expectancy compared with some other regions of the world (Vos et al., 2020). Compounding these problems, many communities, especially those in rural areas, are exposed to heavy pollution and lack access to specialized medical care. Overall, there is a lack of research infrastructure necessary to monitor and respond to these issues.

Fortunately, many of these challenges can be overcome through preventive and public health initiatives. To enact such initiatives, Linkov said, it is imperative to improve Central Asian research infrastructure, knowledge of effective research methods (especially statistics), and project planning and management resources. In addition, she said that researchers in the region could benefit from better resources and strategies for avoiding predatory publishing practices; disseminating their findings; applying for international funding; and accessing technology, biological samples, research data, pathology services, and research participants.

The imperative to attend appropriately to the ethical considerations surrounding data access and security is important in addressing the above challenges, Linkov said. For this, she suggested that a change to the current research culture is needed. In particular, she emphasized the need for scientists to prioritize best practices such as establishing collaborative publishing teams and data banks; teaching research methods, grant writing, and publishing techniques; and adopting common institutional review board frameworks.

Data Governance for Life Science in Uzbekistan

Ravshan Azimov, Uzbekistan Ministry of Higher Education, and Innovation, and Tashkent Medical Academy (Uzbekistan), discussed data governance concepts and approaches relevant to life science research in Uzbekistan.

Proper data governance ensures that data are collected, processed, and transferred in accordance with study protocols and without being changed, concealed, or distorted in any way, Azimov said. As part of data governance, quality control measures ensure effective research, justify scientific funding, and facilitate the generation of study results that can form the basis for important health management decisions that directly affect people's lives. In addition, generally recognized international ethical norms urge that biomedical and behavioral research be conducted with respect for people, mercy, and fairness (CIOMS, 2016).

Although research is distinct from the practice of medicine and public health, efforts to collect, analyze, and interpret biomedical data are critical to improving human health, Azimov said. Uzbekistan has introduced new measures to finance research projects, improve staff training and project management, and foster a modern research infrastructure and information environment.[1] In addition, Uzbekistan has enacted targeted measures, laws, and research standards that will improve science effectiveness, make scientific data more widely available, develop interdisciplinary research, and build a knowledge infrastructure with the aim of spurring innovation and economic development. These measures include mechanisms for improved collaboration, digital dissemination, secure data validation and storage, quality controls, and compliance with international ethical and legal measures. Full enactment of these measures will require financial resources and responsible, careful implementation of good practices and standards, Azimov said.

Cancer Genetics and Genomics Studies in Kazakhstan

Gulnur Zhunussova, Institute of Genetics and Physiology (Kazakhstan), highlighted research efforts in cancer genetics in Kazakhstan. Her laboratory at the Institute is one of several scientific institutions with programs in cancer genetics and oncology. As an example, Zhunussova described a project for which she conducted a comprehensive search for genetic predisposition to early-onset breast cancer using next-generation sequencing, and identified novel pathogenic variants that could contribute to the high rates of this cancer seen in Kazakhstan. She has also studied molecular and genetic principles relevant to diagnosis and screening for colorectal cancer, drug development for early diagnosis and targeted therapy of breast cancer and prostate cancer, epigenetic testing for the diagnosis of colorectal cancer, and new molecular genetic methods for preclinical diagnosis of aggressive forms of prostate cancer (Zhunussova et al., 2019).

While some of Kazakhstan's scientists can access sophisticated laboratory equipment and several genetic and genomic databases, Zhunussova noted that an important limitation is that the population of Central Asia is not well-represented in most databases. In addition, she said, the region's scientists sometimes struggle to find adequate funding to support large comparison studies.

---

[1] Law of the Republic of Uzbekistan; LRU 630-Сон 24.07.2020 (see https://lex.uz/docs/5155423).

Alzheimer's Disease Research in Kazakhstan

Sholpan Askarova, Nazarbayev University (Kazakhstan), discussed progress and challenges in the study of Alzheimer's disease in Kazakhstan. The etiology of this disease, characterized by beta amyloid plaques in the brain that cause progressive loss of memory and abilities in older adults, is unclear and multifactorial. By investigating gut microbiota alterations in Central Asian patients with Alzheimer's disease, researchers found correlations between disease severity, certain bacterial taxa, and biochemical blood parameters (Askarova et al., 2020; Kaiyrlykyzy, et al., 2022a,b). This research followed ethical data consent, collection, storage, and access guidelines based on those of the U.S. National Institutes of Health, and the results were shared with the global research community.

While this research represents an important step forward, Askarova noted that her team has faced substantial barriers in pursuing Alzheimer's disease research in Kazakhstan. She said that the disease is stigmatized and not seen as an urgent medical problem in Kazakhstan. The country lacked official Alzheimer's disease protocols until 2015, and there is a dearth of trained neurologists to diagnose and treat patients, and a lack of validated neuropsychological tests in the Kazakh language. This has led to a striking lack of fundamental data about the disease's prevalence in Central Asia. In addition, there is no nationwide patient recruitment structure in Kazakhstan, and Central Asia lacks specialized Alzheimer's disease clinics and biobank registries; as a result, few publications or data sources are available for comparisons. Because of these barriers, Askarova said that it took her 5 years to gather enough participants for her studies. However, she remains committed to advancing this work, and her team is now initiating genome sequencing studies and metagenomic bacterial studies.

Antibiotic Resistance Issues in Kyrgyzstan

Kalysbek Kydyshov, Republic-Level Center for Quarantine and Highly Dangerous Infections, Ministry of Health of Kyrgyzstan, presenting on behalf of Nurbolot Usenbaev, reviewed research aimed at addressing microbial threats and antimicrobial resistance in Kyrgyzstan.

Studies of hospital patients have shown several antibiotic-resistant pathogens present in the country, including strains of *Escherichia coli*, *Staphylococcus*, and *Proteus*. Kyrgyzstan currently monitors antibiotic resistance by following the recommendations of both its Ministry of Health and the European Committee on Antimicrobial Susceptibility Testing (EUCAST, n.d.). To improve early laboratory diagnosis of antibiotic-resistant pathogens in the country, researchers have developed an atlas of basic bacteriology; created standard operating procedures for sampling, storage, and transport; and started a collection of research data. Development of a clinical protocol to study microbiota is also underway.

To continue this progress, Kydyshov identified several key needs. In the research realm, he suggested the country needs to increase laboratory capacity, reduce research costs, and investigate genetic mechanisms and resistance genes. Where research and clinical areas intersect, he said, it is important to improve training among clinicians, epidemiologists, clinical pharmacologists, microbiologists, and other specialists; update regulatory and methodological frameworks to determine microorganisms' sensitivity to antibiotics; widely share antimicrobial resistance information and monitoring tools; and develop effective treatments that reduce the risk of complications.

Determination of Bovine Brucellosis Genotypes in Kyrgyzstan

Jailobek Orozov, A. Duisheev Kyrgyz Research Institute of Veterinary Science (Kyrgyzstan), shared research on brucellosis, an infectious disease that can be transferred to people from farm animals such as sheep and cattle. Brucellosis is an underdiagnosed and underreported public health and veterinary medicine problem in Kyrgyzstan, Orozov stated, although human infections have plummeted since a national animal vaccination program began in 2011.

Orozov explained that brucellosis infections can be studied through epidemiological monitoring; serological methods; bacteriological methods; and molecular-biological methods, such as gene sequencing and polymerase chain reaction tests. Farm animal movements are registered and monitored in Kyrgyzstan via the country's Animal Identification and Tracking System, which reinforces national oversight, provides systems to prevent and respond to infections, and encourages global and domestic cooperation.

In addition to domestic genetic research, Orozov noted that scientists in Kyrgyzstan also collaborate with researchers in other countries, including by exporting samples, through memoranda of cooperation in line with the Cartagena and Nagoya Protocols (Convention on Biological Diversity, n.d.a,b). These memoranda are amended from existing templates to fit the conditions of the research, Orozov noted.

Research on Flora and Fauna in Kazakhstan

Margarita Ishmuratova, Buketov Karaganda University (Kazakhstan), highlighted goals and challenges of studying biodiversity in Kazakhstan, which is the world's ninth-largest country and home to many unique endemic species. Plant and animal biodiversity is vitally important to global ecosystem stability, yet it faces serious threats from anthropogenic activities and climate change. Unfortunately, Ishmuratova said, a variety of factors create barriers to cataloging and studying biodiversity in Kazakhstan. Some of these factors are environmental; for example, the country's desert climate makes field work challenging. In addition, there is a lack of digitization of scientific resources. The country's scientific journals are neither digitized nor openly accessible (Ryabushkina et al., 2016), and because most herbarium collections are not present in electronic form, researchers can only access samples and data in person, Ishmuratova said. She continued, "That means researchers, especially young beginners, have a problem—we do not have normal access to information and foreign colleagues don't have that access either."

Even where data from Kazakhstan are digitized, such as in international databases like the Global Biodiversity Information Facility (GBIF), she noted that it is not representative of the entire country (GBIF, n.d.a). To ensure better access to Kazakhstan's biological diversity data, scientists from across the country are collaborating to conduct workshops aimed at teaching researchers and students how to digitize and disseminate biodiversity data through GBIF and other platforms.

SUMMARY

As several speakers noted, Central Asia's life science data are of great value and deserve to be both accessible and well protected. This first workshop offered a view of the life science research landscape in select Central Asian countries, covering a wide range of areas from chronic diseases to antimicrobial resistance to animal diseases and biodiversity.

Drawing from the experiences of researchers and data experts from Kazakhstan, Kyrgyzstan, the United States, and Uzbekistan, workshop participants also shared examples of important research infrastructure needs and considerations for future data practices and norms, data management strategies, and data governance tools in the region, laying a foundation for deeper discussions of these areas in future workshops.

2

Data Governance Principles for Life Science Research Across the Globe

Lusine Poghosyan, Columbia University (United States), and Trisha Tucholski, U.S. National Academies of Sciences, Engineering, and Medicine, introduced the second workshop, which was held on May 11, 2023, and focused on data governance principles for life science research across the globe. Tucholski noted that the rapid increase in life science data presents tremendous opportunities for research insights; these data represent significant financial investments and are sometimes sensitive. She stated that global research collaborations seek to make the best use of life science data to advance science and public health, yet there are challenges in balancing the need for data access and sharing with the imperative to ensure responsible and reliable data protections for those data and the people who contribute them.

To explore these issues in a governance context, participants in the second workshop highlighted global data governance policies, practices, and frameworks; examined international norms and emerging trends in life science data governance; and considered similarities, differences, challenges, and opportunities in existing data governance frameworks.

DATA GOVERNANCE PRINCIPLES FOR LIFE SCIENCE

Yann Joly, McGill University (Canada), introduced global data governance resources and frameworks to set the stage for the workshop discussions. Data governance includes agreed-upon data sharing and interoperability standards that promote open science and global collaboration, accelerate health research, and improve health care, he said. Joly also noted that such governance requires an infrastructure of national and international laws, regulations, and standards that address the ethical, legal, and social issues raised by data collection, such as consent and confidentiality, while ensuring data access.

Joly identified several key organizations and texts that promote open science and facilitate responsible international data sharing. These include:

- The United Nations Educational, Scientific and Cultural Organization's *Recommendation on Science and Scientific Researchers* and *Recommendation on Open Science* proclaim the right for people to benefit from scientific advancement, which is a foundation of open science (UNESCO, 2017, 2023).
- The World Health Organization's *Sharing and Reuse of Health-Related Data for Research Purposes* is a guide to data sharing and protection, which was adopted in 2022 in part to capture lessons learned from data sharing and sample sharing experiences during the COVID-19 pandemic (WHO, 2022b).
- The Convention on Biological Diversity's Nagoya Protocol establishes procedures for international sharing of genetic data; Joly noted, however, that it is generally understood to apply only to nonhuman biological samples and has not been ratified by the United States (Convention on Biological Diversity, n.d.b).

- The FAIR (Findability, Accessibility, Interoperability, Reusability) Guiding Principles is a framework that has become a community norm for scientific data management and stewardship, as computational systems handle increasingly large and complex collections of life science data (GO FAIR, n.d.).
- The Global Alliance for Genomics and Health has several products to facilitate responsible access to life science data, including tools for finding data and interpreting the consent associated with them, and a framework for responsible data sharing (GA4GH, n.d.).
- The Public Health Alliance for Genomic Epidemiology promotes ethical data sharing for nonhuman microbe and pathogen specimens to respond to disease outbreaks (PHA4GE, n.d.).
- The Global Initiative on Sharing All Influenza Data promotes rapid sharing of influenza and coronavirus genomic data, but it has been controversial and faces an uncertain future (Re3data.org, n.d.).
- The European Union's General Data Protection Regulation (GDPR) is a legal framework for the protection of individual privacy that applies within Europe as well as to any situation in which data are shared with people or organizations in Europe (European Commission, n.d.).

This list, while not comprehensive, illustrates the breadth of the efforts to guide data sharing at the international level and some of the common elements among them. In general, Joly said that international data sharing practices are guided by broadly formulated, flexible, and easily adopted community norms and standards. While there is broad agreement on the importance of data sharing in science, he noted that challenges can arise when national privacy laws differ from these norms or when scientists struggle to adhere to them. The implementation and enforcement of these norms and standards occurs largely through voluntary adoption and through their influence on national laws, rather than through international laws, which tend to be difficult to enforce. Finally, he noted that most efforts in this area have been dominated by organizations in North America and Europe, but that is gradually changing as organizations on other continents have become more active in deliberations and activities around data sharing.

DATA GOVERNANCE ACROSS THE GLOBE

John Ure, Access Partnership (Singapore), introduced a discussion among data governance experts from around the world. Panelists from Germany, India, Taiwan, Uganda, and the United States highlighted national data governance laws, tools, and frameworks in their respective countries and explored emerging trends in global life science data more broadly.

U.S. Federal Policies for Managing and Sharing Research Data

Taunton Paine, U.S. National Institutes of Health (NIH), discussed how NIH and other federal agencies in the United States have approached data sharing. The U.S. government declared 2023 the "Year of Open Science," drawing attention to the goal of broadening access to life science data. Increased data access and sharing accelerates research by enabling the validation of published research, making high-value datasets more accessible for reuse, and enhancing the rigor and reproducibility of publicly funded research. With appropriate mechanisms, it also increases scientists' opportunities for citation, recognition, and collaboration. Data sharing also fosters transparency and accountability, demonstrates good stewardship of taxpayer funds, and maximizes the value gained from research participants' contributions, all of which promote public trust in research (Funk et al., 2019).

Despite the widely recognized benefits of open science, most of the data underlying published research are currently difficult to access (Errington et al., 2021; Gabelica et al., 2022; Tedersoo et al., 2021). In an effort to close this gap, NIH's primary contribution to the Year of Open Science is its 2023 Data Management and Sharing Policy, a 10-year, multicommunity effort aimed at improving data sharing (NIH, n.d.). Consistent with prior policies, such as the 2014 Genomic Data Sharing Policy, this policy provides supplemental information on sharing human data via a controlled-access framework with appropriate consent, oversight, review, and approval for future use (NIH, n.d.). The policy makes NIH funding contingent on the provision of detailed data management and sharing plans, requiring that researchers address the type of data to be collected and generated; identify the specific tools, software, documentation, or data standards that will be used; provide details about the specific data repository that will be used; outline how and when the data will be findable, protected, and shareable; and delineate future limitations on the use of the data. The policy also prioritizes established data repositories as the preferred method for sharing and requires that data sharing be an early component of research design, not an afterthought, Paine said. While these are not legal requirements, noncompliance can affect funding decisions, which generally exert a strong influence on scientific practice.

Paine also noted that the White House Office of Science and Technology Policy issued a memo requesting that other federal agencies implement policies like NIH's Data Management and Sharing Policy, suggesting that such requirements and practices are likely to continue to gain steam in the United States in the near future.

Life Science and Biological Data Sharing under the General Data Protection Regulation

Fruzsina Molnár-Gábor, Heidelberg University (Germany), described the data governance framework within the European Union (EU) General Data Protection Regulation. As an EU regulation, the GDPR generally applies in all EU member states without requiring them to take additional legal or regulatory measures to apply it in those countries. She highlighted provisions of the GDPR relevant to the processing of life science and health-related data that have been subject to different interpretations.

The GDPR generally prohibits the processing of personal data, Molnár-Gábor stated. However, as there is no absolute definition of personal data, and given that rights and interests in data protection must be balanced with the rights and interests in data processing, exemptions are granted depending on the context in which the data are processed. Assessing risks to subjects from data processing, she said, particularly the risk of reidentification, is crucial. The idea of anonymized data is often understood differently in different technical and legal contexts, which can complicate research, especially in the context of metadata processing, publication, and possible reidentification through artificial intelligence. In addition, the GDPR requires that data processing must be legally authorized, such as through the data subject's consent, including for future and retrospective use. An exception to the general prohibition on processing of sensitive data based on balancing of research interests with data protections is permitted, Molnár-Gábor noted.

Data subjects can consent to the collection and use of their data, and under the GDPR, data subjects have rights that build on the transparency of data processing, such as the right to provide access, delay access, or rectify data. Different roles exist for those who process data, such as controllers, who decide on the purpose and essential means of data processing, and processers, who carry out data processing. In health research, Molnár-Gábor noted, these roles are often shared and exercised jointly, which is acceptable under the GDPR if issues of liability are considered. When research organizations act as data controllers, they are responsible for ensuring the rights of data subjects.

Transferring data outside of the EU presents challenges, said Molnár-Gábor. Under the GDPR, international data transfers require a two-step process that first identifies a legal mechanism for the transfer of data. Next, if the receiving country does not provide protection for the data equivalent to the level of protection provided by the GDPR, data exporters must assess the country's level of protection and compensate for the lack of equivalence with a transfer mechanism that allows for data exchanges. Secure data transfers could be solved with improved regulatory interoperability between countries, Molnár-Gábor said, although this poses further challenges for regulators.

Finally, Molnár-Gábor said that other developments in EU law, such as the proposed 2023 European Health Data Space regulation (European Commission, n.d.) for the processing of health care and health research data, could simplify the sharing of life sciences data with non-EU countries but also create new compliance challenges. Molnár-Gábor added that while there may be issues with interpretating black-letter laws, the documenting, monitoring, and enforcing of GDPR principles can strengthen public trust by promoting transparency and accountability and empowering research subjects.

Data Governance, Principles, and Structures for Life Science and Medical Research in India

Athira P.S., National University of Advanced Legal Studies (India) discussed steps India is taking to address what she described as the country's "growing pains" that have accompanied its digital revolution. As many aspects of Indian life have come online, from banking to education to health care, P.S. said that numerous efforts have arisen to provide a vision for governance in line with principles of security, trust, and digital empowerment (India Stack, n.d.). While India currently relies on guidance and regulations, not legislation, to ensure proper data management, she said that as technology advances, there is an increasing need for more inclusive, clear, secure, sustainable, and cohesive structural mandates for data governance that stress personal privacy.

One effort in this space is the proposed federated structure Data Empowerment and Protection Architecture (DEPA), which P.S. described as a collection of legal protections for personal health data that balances portability and access with security and public trust and represents a departure from the GDPR model of regulation (NITI Aayog, 2020). Under DEPA, she explained, each Indian citizen would be given a unique health identifier, linked through multiple channels, that would be protected and anonymized, yet accessible anywhere in the country. Along with the Digital Information Security in Healthcare bill (which also has not yet been enacted), DEPA would bring health data storing and sharing standards in India in line with international standards and protect research participants' fundamental rights to autonomy and privacy. These fundamental rights are granted to Indian citizens by the Puttaswamy decision of the Indian Supreme Court (Puttaswamy, n.d.), a decision that also asserted that personal data should be collected for a specific stated purpose, relevant for that purpose, and limited to what is required to achieve that purpose.

P.S. pointed to another relevant mandate under consideration: India's Digital Data Protection Act of 2022,[1] which would cover all digitized personal data, including data collected offline that is later digitized. She said that this legislation includes a concept of "deep consent," which indicates that personal data can be used without express consent if such use is necessary for national security, public interest, or public order. This legislation also claims extraterritorial operation outside of India; under this framework, P.S. said, approved countries would be considered safe data transfer locations, while transfers to other countries would be prohibited, making legal compliance and data interoperability important concerns.

Regulatory and Governance Frameworks in Human Biological Materials and Data Sharing in Uganda's National Biorepository

Hellen Nansumba, Ministry of Health of the Republic of Uganda, described existing frameworks for data sharing in her country, along with some of the challenges involved in strengthening equitable data sharing practices. Genomics is a fast-growing field in Uganda, with the scope of work expanding and genomic surveillance accelerating after the COVID-19 pandemic. Nansumba explained that the country is instituting several frameworks for sharing human biological materials and associated data that include ethical requirements, such as informed consent, confidentiality, and impartiality, as well as governance requirements, such as biorepository management and transfer agreements for materials and data.

Uganda's National Biorepository, which stores samples and data from several researchers and institutions, requires anyone depositing or accessing data to follow a specific workflow and requirements that are set at the point of funding, Nansumba explained. The data are first uploaded into the National Health Laboratory Information Management System, then undergo quality processing, and are then transferred to the biorepository and stored appropriately. To access data from the biorepository, researchers must request approval through the appropriate national Research Ethics Committee, satisfy the data and material transfer agreements, and upload the resulting findings and data back into the repository. "We expect [a] return of research findings, for example, in genotypes and microbiome data," said Nansumba.

Nansumba suggested that Uganda and other low- and middle-income countries need stronger infrastructure and technology transfer capabilities to participate in more open, transparent, and equitable global data sharing among researchers, health systems, and patients. She noted that many countries do not have the focused information management systems required to store and transfer large amounts of genomic data. Nansumba suggested that collaborators establish a plan to share the costs of sustaining data over long periods of time, after a project has ceased. She also suggested that sharing research findings with

---

[1] Since the workshop, the Digital Data Protection Act was enacted in August 2023 (see https://www.meity.gov.in/writereaddata/files/Digital%20Personal%20Data%20Protection%20Act%202023.pdf).

patients will be important for improving transparency and strengthening equitable data sharing practices. Finally, Nansumba pointed to a need for building technology transfer capacity to answer questions related to the availability of data for vaccine and pharmaceutical development.

Taiwan's National Health Insurance Research Database

Ya-Hsin Li, Chung-Shan Medical University (Taiwan) discussed governance approaches used for data in Taiwan's National Health Insurance Research Database. Taiwan's national health insurance, paid for by the government, has covered most of the Taiwanese population since 1995, Li explained, making the National Health Insurance Research Database a rich resource for studying real-world health data. This database links more than 400 sources of data covering nearly everyone in Taiwan. It includes health data from cancer registries, cause-of-death records, and many other sources; social data, such as information on factors like smoking or aging; and welfare data, such as information on single-parent families or sexual assault. It can also be used to analyze trends for diseases and treatments or conduct cross-national database comparisons on specific topics.

In Taiwan all the data are encrypted, deidentified, shared, and controlled by the Ministry of Health and Welfare, Li said. Data are linked with demographic variables, such as location, sex, age, diagnoses, prescriptions, and care visit details, but not laboratory tests or medical notes. Applications to work with national health data must be approved by the Ethical Review Board and data can only be accessed within the physical Data Science Centers, which are present in each major city. Li said that the requirement to physically visit a data center to gain access to its data is a strength in terms of privacy protection and works well in Taiwan, but noted that such an arrangement might be less feasible in larger countries.

APPROACHES TO DATA GOVERNANCE

Wei Zheng, Vanderbilt University (United States), moderated a discussion following the panelists' remarks. Panelists were asked to consider similarities and differences among the laws and policies presented and discuss gaps and challenges associated with creating data governance policies. The discussion centered on data governance challenges and consent issues.

Data Governance Challenges

Panelists reflected on some of the unique attributes and main challenges for data governance in different countries. The NIH has demonstrated leadership in prioritizing data sharing in controlled-access public repositories, Paine stated, but it still faces challenges in balancing open access and sharing with participant autonomy, informed consent, and privacy. Another challenge is obtaining the resources needed to support data access and sharing. "Sharing data is not free, and it does require resources," Paine said. To address this, NIH has invested significant resources in infrastructure, such as databases for genotyping and phenotyping, and asks researchers to include data management and sharing costs in their grant proposal budgets. Nevertheless, Paine said that cost will remain an important consideration going forward. Finally, Paine stated that providing access to large, complex genomic datasets can pose technical challenges that must be addressed.

Molnár-Gábor identified three key challenges in implementing the GDPR. First, it is not easy to determine exactly which sector-specific data and privacy measures for processing health data are GDPR compliant. Often, this task falls to researchers or data processors, who could benefit from concrete rules, rather than the existing, more general provisions. Second, she noted that it is difficult to translate legal requirements into technical data protection measures. Lastly, many legal mechanisms that allow data transfer outside the EU/European Economic Area require, on the one hand, data exporters to conduct a comprehensive assessment of the level of data protection in the country receiving the data and, on the other hand, data importers to sign up to binding rules that complement their national data protection regulations, making international collaborations more difficult overall, Molnár-Gábor said.

Taiwan's biggest challenge, Li stated, is ensuring when truly informed patient consent for research purposes has been collected, as much of its national data comes from routine health visits. Nansumba identified Uganda's biggest challenge as a lack of transparency around the regulatory requirements for data access, especially for secondary use.

Broad Consent

Joly asked about the practice of gaining broad consent, which can eliminate the need to reestablish consent when new research is conducted using existing data. Paine replied that laws vary in the United States, but for genomic data, NIH's consent framework does cover future use, with potential limitations depending on what the data are and how consent was structured. In the EU, Molnár-Gábor said that the definition of broad consent, as well as detailed rules for its application, are not included in the main body of the GDPR, which creates the potential for different definitions and rules for application in member states. In general, she said, transparent compliance with informational obligations is necessary, regardless of consent as the legal basis for data processing, to enable data subjects to enforce their data protection rights.

Damira Ashiralieva, National Scientific-Practical Center, Ministry of Health of Kyrgyzstan, noted that, in her country, informed consent is taken before sampling, a practice that proved helpful when COVID-19 diagnostics were used for further research. However, she said, the country's national legislation still needs to be fully harmonized with international requirements. Nansumba stated that Uganda's national biorepository does use broad informed consent, because most of the samples are from routine health care visits and are used for future research, and researchers do share their findings with interested participants. Clinicians also specify when samples will be used in the future—for example, to study hereditary diseases or for public health initiatives.

SUMMARY

In closing, Tucholski underscored the challenges that researchers, organizations, and countries face in balancing open access to data, data interoperability, and individual privacy rights. The data involved in life science research endeavors are often sensitive, and safeguards are likely important to protect them while they are stored, shared, transferred, and accessed, especially across national borders. She noted that subsequent workshops in the series would further explore benefits and risks involved in sharing data, as well as considerations for achieving equitable data sharing.

3
Opportunities and Challenges for Life Science Data Sharing

The third workshop in the series, held on June 1, 2023, delved into issues of ethics and equity in the context of data governance and stewardship. Trisha Tucholski, U.S. National Academies of Sciences, Engineering, and Medicine, and Lusine Poghosyan, Columbia University (United States), provided opening remarks. Building upon the first two workshops, they reiterated that the increasingly large, digitized datasets generated and analyzed by life science researchers present tremendous opportunities for international collaborations to advance innovation and address global health challenges, but such opportunities also come with the responsibility to share data securely and appropriately to protect individual privacy; and government, institutional, private investments; and national security. The third workshop was focused on examining the benefits, risks, and vulnerabilities involved in data sharing; best practices for feasible and equitable data sharing; and ways to navigate policies for authorized access to data while preventing unauthorized access.

INDIGENOUS KNOWLEDGE, BIOLOGICAL STEWARDSHIP, AND COMMUNITY DATA GOVERNANCE

To open the workshop and elicit some of the ethical and equity dimensions of data governance, Krystal Tsosie, Arizona State University (United States), provided a perspective on the complex relationship between Indigenous knowledge systems and Western science in the context of data ownership and stewardship.[1] Historically, she said, scientific research has been rooted in the colonial act of "discovering," collecting, and displaying specimens in museums. However, she characterized these as acts of biopiracy, because while the discoveries are attributed to Western scientists, the knowledge often originated with Indigenous peoples (Das and Lowe, 2018; Davis, n.d.).

Agriculture provides several examples. Indigenous agricultural systems were considered "primitive" by settlers, yet most of today's cash crops and medicines are derived from colonized land and knowledge. Furthermore, an estimated 30,000 edible plants flourished worldwide during the precolonial period, but the combination of industrial farming and the highly mechanized Green Revolution of the 1970s resulted in a drastic loss of biodiversity and widespread ecosystem degradation, and further entrenched societal inequities. Today, a mere 30 plant species constitute most diets worldwide, and Indigenous agricultural practices are seen as the key to the sustainability efforts needed to rescue degraded ecosystems, Tsosie noted. Heirloom seeds and varietals long stewarded by Indigenous people are being sought to reintroduce biodiversity, yet Tsosie cautioned that new gene editing

[1] According to the National Library of Medicine Data Glossary, *data ownership* refers to the "legal control of and responsibility for data," whereas *data stewardship* "involves ensuring effective control and use of data assets and can include creating and managing metadata, applying standards, managing data quality and integrity, and additional data governance activities related to data curation" (see https://www.nnlm.gov/guides/data-glossary/data-ownership and https://www.nnlm.gov/guides/data-glossary/data-stewardship).

techniques threaten, once again, to enable Western scientists and companies to benefit unilaterally from Indigenous knowledge.

Tsosie argued that the concept of "open data," which purports that data should be freely accessible to advance life science discoveries for the benefit of all, ignores the fact that Indigenous communities have struggled for centuries with unethical research practices, strained relationships with settlers, and trust eroded by their descendants. Rather than generating benefits for all, these communities have learned from experience that research benefits often flow in only one direction: out of their communities. In one example, scientists developed an antimalarial compound based on information they had gained from Indigenous communities in French Guiana, without attributing that knowledge to the community members in the patent the researchers received (Pain, 2016). In response to their history of such experiences, Tsosie said that some Indigenous communities are now restricting data sharing, and in some cases even forbidding genomic research using biodata originating in their communities.

It is important to recognize that patent systems heavily favor Western forms of intellectual property, Tsosie said. Indigenous practices are often based on generations of stewardship, experiential learning, and hypothesis generation, practices that are not typically granted patents. In fact, the very concept of ownership is often in conflict with Indigenous practices. However, even if Indigenous communities attempt to protect the intellectual property they generate, they face significant inequities in accessing the legal resources to do so, she said. Universities and corporations have powerful legal and financial resources, including specific legal frameworks protecting the right to patent inventions that result from federally funded research at U.S. universities, but most Indigenous communities—which may not even be legally recognized by the federal government—do not. Furthermore, she noted that companies and large research organizations are disincentivized from using knowledge and genomic data from Indigenous and other minority populations to specifically benefit those populations because of the structural context in which discoveries are monetized, which runs counter to equity aims. "When you tie innovation to economic activities, then we are always fundamentally going to disenfranchise the few and the minority, and this is not a definition of equity that we need to be advocating for," Tsosie said.

Increasing attention is now being focused on the importance of equitable research practices to ensure that benefits are shared with those who contributed data, continued Tsosie. The Nagoya Protocol, a legal framework under the Convention on Biological Diversity, requires fair and equitable benefits sharing. However, the Nagoya Protocol does not provide a detailed roadmap on how to achieve this, nor is the United States a signatory to it, meaning that tribal nations within the territory of the United States cannot operationalize its stipulations or engage in global collaborations. What these communities need, Tsosie said, is intrinsically defined rights to exercise autonomy and protect their genomic data, a concept she called Indigenous genomic data sovereignty.

Tsosie said that one outcome of the growing movement to counter Indigenous disenfranchisement in science is the Native BioData Consortium (n.d.), launched with an aim of empowering Indigenous communities with data sovereignty.[2] It is a nonprofit,

[2] As defined by the National Library of Medicine Data Glossary, "data sovereignty refers to a group's or individual's right to control and maintain their own data, which includes the collection, storage, and interpretation of data. Indigenous data sovereignty refers to the ability for Indigenous peoples to

Indigenous-led biological data repository that uses digital tools and machine learning approaches, such as blockchain technology and federated learning, to protect Indigenous people's control of their genomic data and encourage equitable, beneficial research (Boscarino et al., 2022; Mackey et al., 2022). Another example, shared Tsosie, called Local Contexts (n.d.), is a global initiative that provides Indigenous communities with tools for defining data attribution,[3] access, and use rights to support data provenance[4] and enable greater transparency and integrity in research.

The status quo in scientific practice, and the growing movement toward open science, privileges researchers' access to data, but Tsosie cautioned that it fails to adequately attend to the equity implications of data access and sharing. She argued that researchers have a responsibility to advance equity not only through the recruitment of study participants, but also through the inclusion of communities as partners in knowledge generation. This means incorporating benefits and ethics into the full data cycle and including community voices at all stages of research, not just at the end or as a means of starting a study (McCartney et al., 2023). She also asserted that communities who participate in research should see benefits of that participation in the near term and not just as some distant outcome that may or may not emerge after a scientific paper is published.

In contrast to FAIR (Findability, Accessibility, Interoperability, Reusability) data principles, which do not explicitly address the inclusion of community members as a part of the research process, Tsosie suggested focusing on CARE data principles, which emphasize Collective benefit, Authority to control,[5] Responsibility, and Ethics (Carroll et al.,

---

control their data and includes autonomy regarding a variety of data types such as oral traditions, DNA/genomics, community health data, etc. Within the context of transnational indigenous sovereignty and self-determination movements, indigenous data sovereignty can be a powerful tool for those whom the data represents, which claims the rights of Indigenous peoples to use and interpret the data in a way that is accurate and appropriate given their circumstances, customs, and communal way of life" (see https://www.nnlm.gov/guides/data-glossary/data-sovereignty).

[3] Generally, *data attribution* refers to crediting or ascribing the source of data.

[4] As defined by the National Library of Medicine Data Glossary, "data provenance, sometimes called data lineage, refers to a documented trail that accounts for the origin of a piece of data and where it has moved from to where it is presently. The purpose of data provenance is to tell researchers the origin, changes to, and details supporting the confidence or validity of research data. The concept of provenance guarantees that data creators are transparent about their work and where it came from and provides a chain of information where data can be tracked as researchers use other researchers' data and adapt it for their own purposes" (see https://www.nnlm.gov/guides/data-glossary/data-provenance).

[5] In this context, *authority to control* refers to Indigenous peoples' authority to control and govern their data. "(The) United Nations Declaration on the Rights of Indigenous Peoples affirms Indigenous Peoples' rights and interests in their data. Recognition of these rights bolsters Indigenous Peoples' authority to control and govern such data, further affirming the need for 'data for governance.' Indigenous Peoples must have access to data that support Indigenous governance and self-determination. Indigenous nations and communities must be the ones to determine data governance protocols, while being actively involved in stewardship decisions for Indigenous data that are held by other entities" (Carroll et al., 2020, p. 6).

2020, p. 3).[6] Describing equity as "both a process and an outcome" (NASEM and NAM, 2023, p. 41-42), she emphasized the importance of upholding Indigenous data sovereignty, fighting existing power dynamics, and ensuring full opportunity and access for Indigenous communities as vital to centering equity in scientific decision-making, engagement, and benefits.

CHALLENGES OF TRADITIONAL KNOWLEDGE PRESERVATION AND PROTECTION IN CENTRAL ASIA

Zhyldyz Tegizbekova, Ala-Too International University (Kyrgyzstan), discussed traditional knowledge in Kyrgyzstan and some of the challenges faced in its preservation and protection. The five Central Asian states are rich in traditional social, cultural, and religious customs, which take many forms, including language, food, medicine, stories, games, arts, crafts, music, dance, and poetry. In reflecting a community's shared values, Tegizbekova said that these customs, which encompass traditional knowledge, distinguish one community from another, provide spiritual meaning, and promote community continuity.

Digitizing traditional knowledge is valuable for knowledge sharing; however, it is also critical to protect traditional knowledge from exploitative commercialization, misuse, or misappropriation, which not only can be culturally offensive but also can cause economic or spiritual damage to a community, Tegizbekova said. She highlighted the example of corporations that may wish to gain traditional knowledge about plants with medicinal properties. "It's not about keeping these traditional herbs in secret, but it's about how benefits will be shared among the medical companies that [learn from] traditional communities, which already use traditional herbs in medicine," she said. Unfortunately, she continued, the existing intellectual property system is insufficient for protecting Central Asian traditional knowledge or for supporting local communities that wish to commercialize their products. Kyrgyzstan has laws preserving and protecting traditional knowledge, as well as an Intellectual Property Digital Library holding more than 1,000 items, but Tegizbekova said that these laws need to be strengthened and their implementation fully funded for them to be effective. She added that the other four Central Asian countries do not have dedicated traditional knowledge legislation, meaning that issues of lawful access to and use of traditional knowledge, as well as genetic resources, are regulated ineffectively.

From a research perspective, Tegizbekova said, traditional knowledge "needs to be recorded and digitalized, but at the same time we have to keep in mind how to respect the IP [intellectual property] rights of Indigenous communities." In addition to specialized legislation, she believes that traditional knowledge protection in the region could be enhanced through awareness campaigns; developing more precise definitions of traditional knowledge and folklore; using expert assistance in collection and digitization; the establishment of national centers for the collection, popularization, protection, and preservation of traditional knowledge; and adopting of regional treaties, laws, and/or programs that support local capacity building for traditional knowledge preservation.

---

[6] Here, *ethics* refers to the centering of "Indigenous Peoples' rights and wellbeing across data ecosystems and throughout data lifecycles in order to minimize harm, maximize benefits, promote justice, and allow for future use. Paramount to ethics in data practices is representation and participation of Indigenous Peoples, who must be the ones to assess benefits, harms, and potential future uses based on community values and ethics" (Carroll et al., 2020, p. 6).

GBIF: KICK-STARTING THE BIODIVERSITY PUBLICATION PROCESS FOR TAJIKISTAN

Samariddin Barotov, Institute of Botany, Plant Physiology, and Genetics of the Tajikistan National Academy of Sciences, discussed how Tajikistan's partnership with the Global Biodiversity Information Facility (GBIF) has enhanced digitization and data sharing efforts among members of the country's biodiversity research community. GBIF is a voluntary, intergovernmental network and research infrastructure that provides free and open access to global biodiversity data (GBIF, n.d.a). It houses more than 2 billion species occurrence records and nearly 80,000 datasets. Researchers in more than 60 countries download 23 billion records per month, and almost 8,000 peer-reviewed papers have been published using its data.

GBIF includes observations, digitized specimens, remote-sensing data, environmental DNA, and other types of data that are linked and shared via common data standards, data indexing, and publishing mechanisms, Barotov said. This makes it a rich resource for biodiversity evidence as to where and when species have lived, information that can be used to guide research goals as well as policies for biodiversity protection. GBIF data have been integral to many publications by universities, museums, governmental agencies and ministries, field scientists, citizen scientists, and businesses.

Barotov explained that, to encourage more publications from Armenia, Belarus, Georgia, Kyrgyzstan, Tajikistan, Ukraine, and Uzbekistan, and also to educate these countries' researchers about open data and data sharing, GBIF Norway launched the BioDATA program in 2018. As part of this effort, the BioDATA Capacity Enhancement Support Program teaches Tajikistani institutions how to best utilize their biological resources, learn about data collection, and publish through GBIF via proscribed steps that include digitization, registration, and data conversion. Through this program, Barotov (2023) and other researchers from Tajikistan published the data of the Herbarium Fund of the Institute of Botany, Plant Physiology, and Genetics of the Tajikistan National Academy of Sciences.

BALANCING RISKS AND BENEFITS

Vasiliki Rahimzadeh, Baylor College of Medicine, moderated a discussion on the benefits, risks, and vulnerabilities of sharing life science data; best practices for sharing data feasibly and equitably; and navigating policies for authorized and unauthorized access.

Law, Ethics, and Ownership

Rahimzadeh asked panelists to discuss how they reconciled the tensions between intellectual property law and Indigenous views of knowledge, data, and ownership. Tegizbekova replied that she believes that most Indigenous communities are happy to share certain knowledge, because—unless it is specialized knowledge that is needed to help a community survive or thrive spiritually—the issue is less one of intellectual property than one of knowledge in the public domain. However, she said, there is a growing awareness of how destructive the sale of products produced by a community or derived from their knowledge can be. In 2019, for example, an Indigenous community in Panama accused Nike of using its design on a shoe, and Nike stopped producing the shoes. The World Intellectual Property Organization is developing an internationally accepted definition of traditional knowledge, which may also help resolve some of the tension, she noted.

Rahimzadeh added that the considerations around human data and intellectual property are different from those surrounding plant or animal data.

Tegizbekova said that approaches to balancing legal and ethical requirements may vary depending on the research, the methodologies employed, and the institution. In general, however, exposing confidential information is a major violation, she said. Damira Ashiralieva, National Scientific-Practical Center, Ministry of Health of Kyrgyzstan, suggested that all biological material should be assessed for sensitivity and accessibility. Ensuring equity and transparency is more difficult to achieve, especially for international collaborations with multiple experts, as there may be national security interests at stake. It is important to weight the benefits against the risks and consequences, which may get even more complicated with the emergence of new technologies, such as artificial intelligence and synthetic biology. Faina Linkov, Duquesne University, agreed that ethical, accessible, and equitable research requires strong regulations, but cautioned that those regulations must be feasible to follow or there is a risk that researchers will simply ignore them.

Equity Considerations

Shalkar Adambekov, Kazakh National University (Kazakhstan), noted that Western research regulations were adopted in response to a long and well-publicized history of unethical experiments, which have contributed to a history and concept of research ethics that Central Asian nations do not necessarily share. He asserted that equity in research is hard to achieve and posited that it should not necessarily be a priority above other priorities in scientific collaborations. He reasoned that international collaborations are inherently unequal because the needs are unequal: Researchers in developing countries need access to resources, such as computing power, and particular areas of expertise, such as biostatistics, while scientists in more-developed countries need data from less-developed countries. In addition, he suggested that Indigenous communities do benefit from these collaborations through the expertise and financial resources that are shared. "Inequality in science might be a good thing, as long as it's done ethically and benefits science," he stated.

Adambekov drew a distinction between life science data and traditional knowledge, which is much harder to understand, quantify, own, and share. Globalization and industrialization have not benefited every country equally, and he suggested that developing nations should balance their desire to protect their cultural identities against the benefits that can be gained through access to new practices and resources. Sharing data or knowledge does not necessarily result in significant loss, nor does it lead to immediate equity, he stated.

Tegizbekova noted that for communities whose rights are weakly protected at the national level, an internationally accepted definition of traditional knowledge will strengthen justice and better protect the rights and culture of Indigenous communities from unethical exploitation by multinational corporations, where priorities often do not align with those of Indigenous communities. Expressing hope that justice could be achieved, she said developing countries must work together to promote benefit sharing, protect genetic resources, and protect the rights of Indigenous communities. Rita Guenther, U.S. National Academies, stated that the type of data being shared or collected—whether large demographic datasets, personal medical data, or data used as building blocks—can influence when and to whom those data should be accessible, and can help determine different approaches for balancing scientific advances with privacy protections and security measures.

Research on individuals, research involving traditional knowledge, and research pursued to generate profits may raise different considerations depending on the perspective one takes, but regardless, she said that it is important to follow key principles of informed consent, proper preparation, and compliance with institutional and governmental regulations that are constructed to balance benefits and risks. However, these concepts can quickly become challenging in practice. Reiterating researchers' responsibility to protect the interests of people and ecosystems, she said that it is important to have honest and nuanced discussions to work through these complex issues and inform better data governance frameworks.

SUMMARY

The third workshop addressed complex issues in the ways different communities contribute to and benefit from life science research and innovation. Speakers highlighted how Indigenous communities in particular have been disenfranchised in data exchanges and often lack the legal and financial resources and expertise to navigate intellectual property and patent law. Participants offered perspectives on how this has played out in the U.S. context and in Central Asian countries, where speakers stressed the importance of access to better resources and expertise, as well as stronger legislation to address traditional knowledge protection and preservation. Speakers suggested that, developing a globally shared definition of traditional knowledge and demonstrating a commitment to more equitable research practices, with different benefit and risk considerations for different data types, could help to address some of these issues.

4
Life Science Data Governance in Practice

The fourth workshop, held on June 8, 2023, focused on eliciting real-world experiences with data governance in the context of life science research in Kazakhstan, Kyrgyzstan, Uzbekistan, and Tajikistan. Additionally, the workshop explored practical approaches to informing data management and governance. Trisha Tucholski, U.S. National Academies of Sciences, Engineering, and Medicine, and Lusine Poghosyan, Columbia University (United States), opened the workshop by reviewing definitions of several key terms and concepts relevant to the workshop series.

Pointing to a recently published paper on the topic, Tucholski defined *data governance* as "the principles, procedures, frameworks, and policies that ensure acceptable and responsible processing of data at each stage of the data life cycle . . . [to] maintain data integrity, quality, availability, accessibility, usability, and security and define data controllership (also known as stewardship)" (Eke et al., 2022, p. 600-601). Each stage of the data life cycle—including data collection, storage, processing, curation, sharing, application and use, and deletion—has unique considerations that impact data governance (Figure 1). For instance, during data *collection*, researchers consider whether informed consent has been obtained to share and use the data, whether biases exist for sampling, and whether there are legal and regulatory differences that affect sharing the data between localities. During data *storage*, researchers consider how long data should be preserved, who owns the rights to the data, who pays for data storage, and how to minimize risks of data leaks as technology changes over time. In the *application and use* of data, researchers consider how to communicate incidental findings that pertain to the health of study participants, how to ensure that the minimal amount of data is used to minimize risk to participants, how

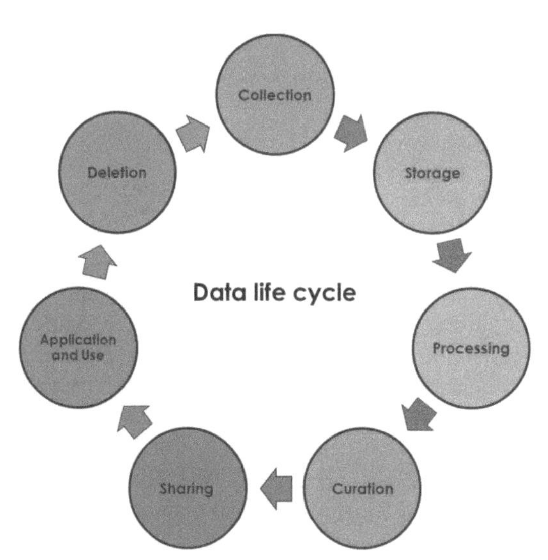

FIGURE 1 The Data Life Cycle. SOURCE: Adapted from content in Eke et al., 2022.

to ensure data are not misused, and how to ensure data are used responsibly for civil and military applications or for economic gain (Eke et al., 2022).

Tucholski also drew a distinction between the concepts of data management and data governance. Data *management* is the technical and practical implementation of data *governance*, which by itself is just documentation, akin to a recipe that results in food through the act of cooking (Everett, 2023). Therefore, to be useful to researchers, data governance policies must be translatable to real-life situations, she said. Frameworks such as the FAIR (Findability, Accessibility, Interoperability, Reusability) guidelines, along with governmental regulations and ethical principles, have been constructed to help ensure equitable access to and benefits from science and innovation.

The workshop aimed to review existing policies, practices, and norms related to sharing and protecting life science data in Kazakhstan, Kyrgyzstan, Uzbekistan, and Tajikistan; discuss their challenges; understand how they translate to real-life field, clinic, and laboratory work; and explore improvements to maximize safe data sharing with multiple stakeholders.

LIFE SCIENCE DATA GOVERNANCE IN CENTRAL ASIA

Tucholski moderated a discussion of the laws and policies applicable to data sharing and protection in Kazakhstan, Kyrgyzstan, Uzbekistan, and Tajikistan.

Ensuring Information Security in Kyrgyzstan

Damira Ashiralieva, National Scientific-Practical Center, Ministry of Health of Kyrgyzstan, discussed the importance of data security broadly, as well as the specific structures that support data security and data governance in Kyrgyzstan. Information security involves several immutable principles, she said, including professional independence, impartiality and objectivity, accuracy and reliability, consistency, clarity and transparency, statistical confidentiality, and relevance. Supporting these principles is central to protecting confidentiality, integrity, and accessibility in science.

The Constitution of Kyrgyzstan and a variety of organizations and laws describe and regulate not only information security but also data governance principles and the right to research, access, receive, and disseminate information. For example, the National Statistical Committee has a communication strategy for improving data dissemination, receipt, and storage. This structure helps to ensure that data are available, authentic, deidentified, and distributed to users in a relevant, timely, objective, accessible, and confidential manner. In addition, the Ministry of Digital Development is launching a database to improve communication, cooperation, data management, efficiency, and information protection for the public, government, and research institutions. The Law on Electronic Governance defines the rights, protections, access, and obligations of the information owner, and the Law on the Conflict of Interest prioritizes public interests and national security, ensures transparency and control, and emphasizes personal responsibility and liability. Other laws cover personal information, biometric registration, and cybersecurity strategies.

Kyrgyzstan, which has a mandatory medical insurance fund, is also creating a modernized, central eHealth hub for its citizens based on the data governance principles of legality, immutability, personalization, verification, and traceability. Access and authorization are treated very carefully within this framework, Ashiralieva said, and research conducted using repository data is coordinated through the National Ethical Committee to rule out conflicts of interest and ensure legal compliance.

Despite these structures, Kyrgyzstan still faces many information security challenges. One important gap is that data are often not available quickly enough to inform decision-making, Ashiralieva said. To address this, she suggested, the country would need more well-qualified staff and improved coordination between the public administration and socioeconomic spheres.

Biological Data Sharing in Uzbekistan

Shakhlo Turdikulova, Ministry of Higher Education, Science, and Innovation and Center for Advanced Technologies (Uzbekistan), described practices and rules involved in biological data sharing in Uzbekistan. As many workshop participants pointed out, technology is generating huge amounts of life science data that hold great potential to advance biology and medicine but also create a need for proper data handling, sharing, and protection, Turdikulova said. Recognizing the importance of life science research to advance the nation's economic prospects, the government of Uzbekistan has made significant investments in scientific centers, universities, and research clinics. Two recent presidential decrees—to improve educational quality and performance in chemistry and biology, and to develop and improve biotechnologies and biological safety systems—also support this work.

Uzbekistan's National Council on Biological Safety is charged with protecting life science data throughout its entire life cycle, and the country is in the process of developing policies, practices, and norms on sharing and protecting biological data. The work has been challenging, Turdikulova noted, but proper planning, related to the collection, storage, sharing, and preservation of research data improves the efficacy, transparency, and dependability of research results. Appropriate forethought is necessary to support regulatory compliance, delineate the responsibilities of involved parties, and ensure that data are securely managed to prevent loss or misuse, she said. In addition, Turdikulova posited that researchers who adopt good data management practices can increase their visibility and influence.

Uzbekistan also prioritizes data sharing, open science, and international collaboration to improve upon its domestic expertise. On the premise that data should be secured, but not restricted, the government of Uzbekistan hopes to establish a Central Asian Genomic Center, an open access platform for plant, animal, and human genomic data like the One Health Initiative of the World Health Organization (WHO).

Looking ahead, Turdikulova identified several key priorities for further enhancing data sharing and data governance in Uzbekistan. First, she said that artificial intelligence techniques are needed to extract insights quickly and accurately from increasingly large and complex datasets; accelerate scientific discovery; and improve the overall understanding of science's most fundamental questions. Incorporating artificial intelligence has been a challenge, but Turdikulova said that the government is committed to supplying the needed specialists, infrastructure, and funds. In addition, she suggested that Uzbekistan needs a more robust digital data management infrastructure to properly store, share, and protect life science data, complemented by stronger regulations and guidelines around data privacy, informed consent, property rights, and ownership. Finally, she said that advancing progress in these areas will require coordination and cooperation among researchers, institutions, and governmental organizations.

Biological Data Sharing in Tajikistan

Zulfiya Davlyatnazarova, Institute of Botany, Plant Physiology, and Genetics of the Tajikistan National Academy of Sciences, described the research environment in Tajikistan and several data protection laws that affect research practices there. The country's Law on Protection of Personal Data; Law on Ensuring Biological Safety, Biological Security, and Biological Protection; and Law on Genetic Resources establish frameworks for data collection, management, and use of individual data, as well as of biological and genomic research, to protect Tajikistan's unique biodiversity resources. The country also has a Healthcare Code for protecting patients' rights.

Scientific research in Tajikistan starts with a proposal, hypothesis, and goal setting; following this, resources, collaborations, and funding are established, and data are collected and analyzed, Davlyatnazarova said. The results are published, typically in Russian and Tajik, with English summaries. Although some data are classified, which confers access limitations, Davlyatnazarova said that most research is open and internationally available and accessible.

International collaborations are an important part of research in Tajikistan, Davlyatnazarova continued, and the country has worked with researchers in China, Japan, and Norway, as well as with the Global Biodiversity Information Facility. Tajikistan shares its genetic data in accordance with national laws, the Nagoya Protocol, and international frameworks to prevent unintended or malicious data use. However, Davlyatnazarova noted, researchers from Tajikistan can find it difficult to obtain the documentation and government approval that international collaborations require. She added that preventing unauthorized access to data, such as taking seeds or genetic materials without permission, is one challenge Tajikistan is attempting to address through clear definitions of what protections certain data types need so that they can be shared securely.

LIFE SCIENCE DATA GOVERNANCE IN PRACTICE

Rita Guenther, U.S. National Academies, moderated a discussion among experts examining how existing policies, practices, and norms translate to real-life scenarios in the field, in clinics, and in laboratories in Central Asia. The panelists—who included Pavel Tarlykov, National Center for Biotechnology (Kazakhstan); Eastwood Leung, International University of Central Asia (Kyrgyzstan); and Elmira Utegenova, Scientific and Practical Centre for Sanitary-Epidemiological Expertise and Monitoring (Kazakhstan)—also explored how existing data governance policies, practices, and norms could be improved or standardized to optimize data sharing.

Tarlykov highlighted the connections between data sharing and the funding mechanisms that support life science research. Like many scientists in Kazakhstan, he depends on research funding from the Kazakhstan Ministry of Science and Higher Education, which requires that research results be published in journals that in turn require data sharing. Furthermore, the national Ethics Committee ensures that proper data management, including deidentification, is in place before funding is granted. If data are collected or shared incorrectly, the research is unpublishable. Tarlykov said that researchers from Kazakhstan commonly use international repositories, such as the National Center for Biotechnology Information and the European Nucleotide Archive, although he noted that Kazakhstan lacks the specialists needed to perform appropriate quality controls.

Utegenova said that in her view, unauthorized access is not a major problem in Kazakhstan. For example, at the National Reference Laboratory for Control of Viral Infections, every sample undergoes ciphering that follows international best practices and protocols. While the effort is ongoing, she said she has not experienced any serious breaches, whether transferring data within Kazakhstan or externally with researchers in other countries or the WHO. She noted that there could be ethical issues if the data are used without referencing the original source, however. She also added that for Kazakhstan to launch a timely, cross-disciplinary response to novel infections, the country needs an improved data exchange platform with appropriate access and sharing protocols.

Leung stated that in Kyrgyzstan, resources are also very limited, and research is expensive. He stressed the importance of establishing consensus on proper procedures before any experimentation begins so that research can proceed in the most efficient way possible. Leung noted that for international collaborations, it is important for the security at every data center to be checked to ensure that the proper jurisdictional procedures and data governance standards are followed, and while work is done increasingly in the cloud, results can also be downloaded onto local servers. Overall, following General Data Protection Regulation stipulations for data transfers would help to address issues of unauthorized access, Leung noted.

SUMMARY

The workshop illuminated a range of existing biological data sharing policies, practices, and protections in Kazakhstan, Kyrgyzstan, Uzbekistan, and Tajikistan. Several participants emphasized the importance of improved coordination, education, infrastructure, and collaboration; described challenges in data availability, expertise acquisition, and unauthorized access; and stressed the importance of following international frameworks and agreements.

5

Examples of Existing and Needed Practices for Cyber- and Data Security in the Life Sciences

The fifth workshop, held June 15, 2023, focused on understanding existing security practices and their associated challenges; discussing needed cyber-, data, and information security practices; and examining differences in the risks and practices across institution types and fields.

CYBER RISK MANAGEMENT IN LIFE SCIENCE RESEARCH

Gautham Venugopalan, Gryphon Scientific (United States), set the stage for the workshop with an overview of cyber risk management in life science research. As defined by the U.S. National Institute of Standards and Technology (n.d.), *cybersecurity* is the "prevention of damage to, protection of, and restoration of computers . . . including information contained therein, to ensure its availability, integrity, authentication, confidentiality, and nonrepudiation." Three foundational principles of cybersecurity are data confidentiality, integrity, and availability (Figure 2). According to Venugopalan, *confidentiality* requires that certain data be kept secret or secure, *integrity* requires that data be accurate and unadulterated to be useful, and *availability* requires that data be accessible to enable normal organizational operations. He said that in open science scenarios, confidentiality is not an issue, but the data must retain its integrity and availability to avoid compromised research, untrustworthy results, or loss of access.

Unfortunately, cybersecurity incidents are increasing, Venugopalan said, and the impacts of these incidents can pose a serious threat to the health and safety of individuals as well as business interests (Check Point Research Team, 2022).

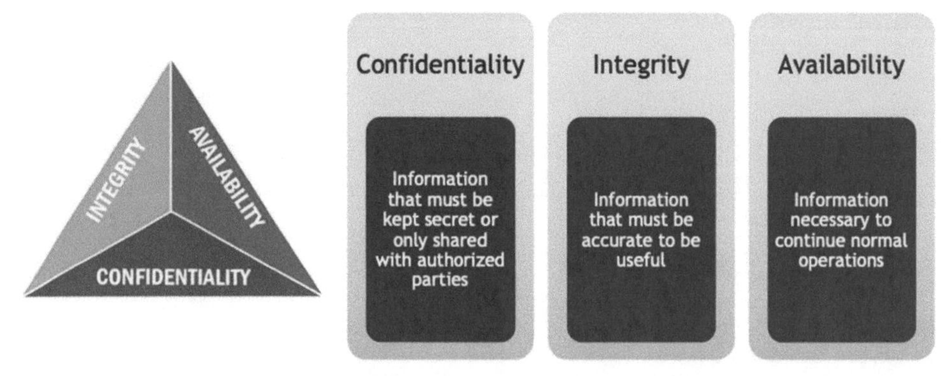

FIGURE 2 Foundational Principles of Cybersecurity. SOURCE: Crawford, E., S. Joshi, C. Garnier, A. Bobrow, N. Tensmeyer, and G. Venugopalan. This research was funded in part by a grant from the United States Department of State (SISNCT21CA3024). The opinions, findings, and conclusions herein are those of the authors and do not necessarily reflect those of the United States Department of State.

High-profile cyberattacks have disrupted pharmaceutical operations and demonstrated the potential to alter vaccine study data or remotely control physical laboratory infrastructure (EMA, 2021; MDL, 2017; Osborne, 2021). Such examples should be alarming to life science researchers, Venugopalan said, as attacks on research assets—from samples to data to internet-connected equipment, including personal smartphones—can result in the loss or delay of scientific advancements, reduced workplace safety, privacy breaches, data misuse, and financial or intellectual property losses. Such attacks can also erode public trust in science and lead to industry-wide disruptions.

To counter these threats and maintain data confidentiality, integrity, and availability, Venugopalan outlined four building blocks of cybersecurity: continuous and iterative cyber hygiene practices, asset identification, scenario building to identify concerns, and control systems for reducing risk (Figure 3). These four practices rely on identifying and protecting assets, detecting and responding to attacks, and creating recovery processes. Together, they can be used to determine an organization's current safety profile, its target profile, and its decision-making needs; recognize and respond to threats and vulnerabilities; train staff; and conduct risk assessments to determine resource allocation.

Cyber hygiene practices are the fundamental controls every organization, no matter its sector or size, must implement before more sophisticated controls can be added, said Venugopalan. These practices, such as staff training, multifactor authentication, data encryption, zero-trust architecture, endpoint device management, and backup strategies (separated and with restricted access), are the most straightforward to implement and protect against a wide variety of common attacks. Quality off-the-shelf cyber hygiene and risk assessment systems exist, such as the Center for Internet Security Controls, version 8. Venugopalan noted that Dropbox and iCloud are commonly used by academic researchers to facilitate data sharing and come with security features, although researchers should agree on appropriate access controls, use, and sharing practices when using such services.

Determining which data or capabilities require the highest level of protection requires collaboration between researchers and information technology staff, Venugopalan said. Unfortunately, he continued, there is no such thing as a foolproof system, because most successful cyberattacks exploit the human factor somewhere along the line. Human behavior can only be improved through training that prioritizes data confidentiality, integrity, and availability, and that creates channels for recognizing attacks. In addition, Venugopalan said that a system's usability is important, because if data access and management systems are too complex or difficult to use, people will find ways to work around them to do their work efficiently. There is no single approach or system that is right for every institution or environment. "An institution has a unique culture, has its own resources, has its own norms," Venugopalan said. "People should work to come up with a system that works for them."

UNDERSTANDING SOCIAL ENGINEERING

Trisha Tucholski, U.S. National Academies of Sciences, Engineering, and Medicine provided a case study examining one common challenge in data security: the human factor. Noting that many of the workshop's participants are on the front lines of life science research and cybersecurity needs, Tucholski said that people are an organization's greatest asset *and* its weakest security link. This is because attackers can manipulate people into

FIGURE 3 An Organization's Cybersecurity Journey. NOTES: The cybersecurity journey typically begins with basic cyber controls and incorporates progressively more intensive processes and protections based on the perceived risks and the resources that are devoted to addressing them. SOURCE: Crawford, E., S. Joshi, C. Garnier, A. Bobrow, N. Tensmeyer, and G. Venugopalan. This research was funded in part by a grant from the United States Department of State (SISNCT21CA3024). The opinions, findings, and conclusions herein are those of the authors and do not necessarily reflect those of the United States Department of State.

divulging information through social engineering, which exploits human psychology, not technology (Dinha, 2023).

The most common form of social engineering is phishing, in which a message prompting staff to share confidential information appears to come from a trusted source. Phishing has become increasingly sophisticated, with attackers using publicly available information to impersonate colleagues or peers (Chhay, 2022). These requests are often worded to create a sense of urgency in hopes that the recipient will overlook errors or discrepancies and act upon the request quickly, giving the attacker access to information or systems that are meant to be kept secure.

Tucholski suggested that training is needed to raise awareness of phishing and to equip staff with the skills to recognize suspicious messages. She noted that unexpected messages from unknown email addresses should be approached carefully, especially if they appear urgent. When in doubt, it is best to validate the sender's identity via another method, such as an independent internet search or trying to reach the alleged sender through other means, such as a phone or video call. In addition, it is critical to avoid sharing log-in information, initiating financial transactions, and clicking on links or downloading attachments from unfamiliar or suspicious email addresses. Even with the best training, however, anyone can be fooled. Attackers are becoming increasingly sophisticated, and increased digitization of everyday life suggests that everyone—but especially staff of laboratories or companies who handle sensitive or proprietary data—be more vigilant in avoiding phishing scams.

DATA AND INFORMATION SECURITY IN CENTRAL ASIA

Kavita Berger, U.S. National Academies, moderated a discussion to examine existing cybersecurity practices and challenges; discuss examples of needed practices for cyber-data, and information security; and investigate differences in risks and practices among institution types and fields. These issues are increasingly relevant for scientists, who could receive phishing attacks disguised as conference registrations, publishing requests, or social media invitations, putting their—and their colleagues'—work at risk. Participants discussed the importance of cybersecurity training, suggested several cybersecurity needs, and discussed the difficulties of balancing data sharing with security and communicating risk to nonscientists. Some participants also commented that the resources mentioned in this workshop can help life science researchers overcome these shared cybersecurity challenges.

The Importance of Cybersecurity Training

Several participants emphasized the importance of continuous training in cyber hygiene practices, risk assessments, and data safeguards to keep everyone involved in life science research—from students to seasoned scientists—vigilant. They noted that data security is still important when working with data that are publicly available. Phishing attacks can be very persistent and penetrate an institution's strongest protections. Venugopalan cautioned that the best way to deal with suspicious or unexpected emails is to delete them immediately. Even just reading an email, without clicking on a link or replying, can open the door to a cyberattack.

Damira Ashiralieva, National Scientific-Practical Center, Ministry of Health of Kyrgyzstan, stated that scientists in her country are becoming more vigilant to the threat of cyberattacks. In fact, she noted that she was suspicious of the invitation for this workshop series and was only satisfied after receiving more information directly from Tucholski. She also noted that the country's public health department is governed by an Ethics Committee that oversees all research and data protections, incorporates expert training, and integrates bioethics to keep up with the rapid pace of science.

Yann Joly, McGill University (Canada), stated that in Canadian universities, cybersecurity expertise to examine the bioinformatic, ethical, and legal components of data is only just being added, and this varies by project and institution. While academic ethics committees exist, they were developed in the context of physical risks, and, he said, they have been slow to incorporate data science expertise.

Cybersecurity is also in the initial stages in Tajikistan, stated Mekhriniso Rustamova of the Tajikistan National Academy of Sciences, which will soon host a cybersecurity conference to educate its scientific workforce. However, Tajikistan's national Bioethics Committee, directly under the Ministry of Public Health, has a well-established, stringent, and successful project approval process.

Cybersecurity Challenges and Opportunities

Wei Zheng, Vanderbilt University Medical Center (United States), suggested that existing systems that ensure verifiable results should be expanded to reduce the risk of data misuse and increase trust in public data repositories. Berger added that public repositories can also lead to data being used in ways that researchers never intended, including ways that may harm people, animals, plants, and the broader environment. Ashiralieva shared this concern, noting that while legislation in Kyrgyzstan and its national Ethics Committee clearly define who can share what information, data used in global collaborations could be accessed by unknown researchers for unknown purposes.

Joly noted that in addition to expanded cybersecurity practices, there is a role for legal contracts in protecting open data and ensuring their integrity. For example, the tools developed by the Global Alliance for Genomics and Health to help scientists share and integrate data safely, legally, and ethically are free and open access but covered by intellectual property licensing to protect them from downstream alteration or abuse.

Finding a Balance

Venugopalan stated that balancing the desire to share and use data openly with the potential for unintended consequences of such sharing and use is less about cybersecurity and more about the values and priorities of laboratories or collaborations and their perception of the risks and benefits. While employing security resources—from cyber hygiene practices to licensing to legal contracts—is important, he noted, that bad actors are unlikely to be deterred. Whether and how to share data is therefore a continuous conversation that can evolve alongside science.

Joly agreed that every research group can make its own choices about how best to protect data at every step, from assembling a dataset to storing and sharing it. He added that circumstances can affect these decisions, noting that the sense of urgency during COVID-19 at times overrode voices suggesting a more cautious examination of potential data inaccuracies and downstream consequences.

Berger said that balancing various risks and benefits of different data types and sharing practices has been discussed in the United States for some time. The term *dual-use research of concern*, for example, describes research conducted for peaceful purposes that could be used by individuals with malicious intent to cause harm. The World Health Organization (2010, 2022a), among many others, offers guidance to help scientists and governments understand the risks and benefits involved in such work. The real challenge, she said, is conducting responsible innovation that recognizes and addresses risks and maximizes benefits, while continuing to build global collaborations and advance science. This challenge is getting more difficult as national and international policies unexpectedly overlapped or counteracted, creating roadblocks for addressing public health emergencies or advancing science. For example, during the pandemic, General Data Protection Regulation restrictions prevented collection of human samples needed for detecting and monitoring the SARS-CoV-2 virus because human genetic material also may be obtained from sequencing those samples, creating concerns about privacy for individuals. She indicated one commonly cited suggestion, which is that global health entities could encourage researchers to secure digital data as carefully and lawfully as they secure physical samples, and added that countries can enhance measures to support open, transparent research, while preventing harms from unauthorized or unlawful access to data.

Communicating Risk to Nonscientists

Rita Guenther, U.S. National Academies, noted that researchers in the United States have struggled to communicate risks to nonscientists, especially decision-makers, who often have very different priorities. One response to a series of biological attacks involving anthrax in 2001 was the creation of the National Science Advisory Board for Biosecurity (NSABB), a group of scientists and academics who lead discussions with government leaders and other researchers on balancing risks and benefits of certain areas of science, such as biosafety and gain-of-function work. She suggested that leaders in Central Asian countries may find NSABB's strategy, which focuses on understanding risk, communicating it, and implementing continuous risk assessment processes, applicable. Ashiralieva agreed and noted that Kyrgyzstan convenes multisector collaborations to address crises such as COVID-19 and other epidemiological emergencies.

Another potential communication model, Berger added, is exemplified by the International Genetically Engineered Machine (iGEM), which integrates risk communication into the broader context of scientific responsibility. iGEM hosts an annual competition for students in synthetic biology, which stresses risk assessment, safety, security, and ethics. In addition, the InterAcademy Partnership and the U.S. National Academies help foreign and domestic researchers investigate dual-use research, uncover biosecurity issues, and embed good security practices, she noted.

SUMMARY

The workshop shed light on the importance of strong cybersecurity practices for life science researchers to combat their shared challenges. An organization's most valuable asset—its people—is also its weakest security link, and participants discussed how students and researchers can be made aware of cybersecurity risks and equipped to employ appropriate practices and protections.

6

Implementing Best Practices for Life Science and Biological Data Governance

In the sixth workshop, held on June 20, 2023, organizers summarized what was learned from the previous five workshops and facilitated a collaborative discussion to explore areas on which to build. They also discussed potential future opportunities to continue the conversation with a broader set of participants. Participants highlighted several concrete suggestions that Central Asian countries can implement to improve their data governance frameworks, infrastructure, and expertise to participate in international collaborations more fully.

HIGHLIGHTS FROM THE WORKSHOP SERIES

Lusine Poghosyan, Columbia University (United States), offered a high-level summary of the previous workshops in the series. In the first workshop, participants presented a sampling of life science research efforts in Central Asia. Scientists from Kazakhstan, Kyrgyzstan, and Uzbekistan, working in veterinary medicine, infectious disease, antimicrobial resistance, genomics, and botany, shared their successes and challenges conducting life science research, digitizing data, and contributing to international databases. She observed some key challenges facing the region's research efforts, with common limitations including a dearth of comparison data from the region; difficulty accessing biobanks and other technological resources; and limited expertise in some areas, such as data science. A key question coming out of these discussions was the degree to which Central Asian governments, institutions, and researchers have identified research goals and investment priorities, Poghosyan noted.

At the second workshop, focused on data governance principles, several participants suggested that successful international collaborations hinge on developing and adopting data interoperability standards and harmonizing national data sharing and security laws. While there are no international data governance laws, several organizations offer resources and guidelines for responsible data sharing across borders, including the United Nations Educational, Scientific and Cultural Organization, the World Health Organization, and the Global Alliance for Genomics and Health.

During the workshop, data governance experts from Germany, India, Taiwan Uganda, and the United States described their data sharing laws and norms as potential models for other countries to learn from in addressing challenges and improving upon their own data governance frameworks. Throughout these discussions, speakers highlighted challenges in data management and international data sharing, such as the cost of long-term data storage and the lack of advanced infrastructure and subject matter expertise in some regions. A question coming out of this workshop was the extent to which data sharing and protection laws in other countries may affect or influence the data governance trends in Central Asia, Poghosyan said.

The third workshop considered the benefits and risks of sharing biological data. While the notion of open science is seen as key to advancing life science research, it is important to remember that tribal and Indigenous communities worldwide have been harmed and disenfranchised by unethical research practices. Through efforts such as the Native BioData Consortium and digital tools such as blockchains, Indigenous communities are exploring methods of protecting and stewarding the data and knowledge their communities hold, in order to have a greater say—and a greater share of the benefits—in research that draws on their knowledge and impacts their communities.

Each of the five Central Asian countries has communities with unique traditional knowledge that not only holds great personal and spiritual meaning but also may be important for the survival of these communities. Several speakers suggested that stronger legal protections may be needed to preserve traditional knowledge. In addition, since there are inequities in communities' ability to access resources and expertise on intellectual property and patent law, empowering these communities to navigate the powers and protections they do have is important. Many participants suggested that different data types require different considerations of risk, use, and sharing, and exchanged different perspectives on the definitions of and roles for ethics, equity, and equality in science. "Through discussion, we discovered that the way we talk and think about ethics, equity, and equality may differ in different contexts and cultures, presenting us with the opportunity to continue having conversations with each other about these essential concepts and the chance to pursue a common lexicon to further facilitate scientific collaboration across borders and across cultures," said Poghosyan.

In the fourth workshop, speakers reviewed existing national practices, policies, and norms around protecting and sharing life science data in Central Asia; described unique features, common challenges, and common goals; and explored ways to improve data governance policies and implementation. Existing policies and practices address electronic governance, conflicts of interest, personal data, biometric registration, genetic resources, and centralized digitization efforts.

During the discussion period, panelists stressed the importance of data sharing with standardized protocols and appropriate security measures. They also described common challenges with data sharing, including limitations in available data and computing power; a lack of expertise in cybersecurity, artificial intelligence, and digitization; and the importance of improved digital infrastructure. Despite these challenges, Central Asian scientists can participate in international collaborations and are establishing a regional genomics repository, Poghosyan noted.

The fifth workshop focused on cybersecurity risk management to maintain the confidentiality, integrity, and availability of life science data. Speakers commented that a commitment to open science includes a responsibility to share data in a secure and meaningful way to advance science. Cyberattacks can disrupt research and compromise sensitive data, and every project can balance open data sharing with risk assessment, especially when international collaborations or dual-use research are involved. Speakers discussed several approaches to advancing cybersecurity in scientific organizations, including basic cyber hygiene, resources, and practices, along with more intensive cyber resilience strategies. Panelists also emphasized the importance of vigilance in the face of social engineering methods, such as phishing, which can manipulate staff into divulging sensitive information, such as passwords and personal information.

KEY LEARNINGS AND REGIONAL CONTEXT

Faina Linkov, Duquesne University (United States), moderated a discussion in which participants reflected on what they learned from the workshop series; discussed differences in approaches, needs, and priorities among countries; and considered how engaging a broader audience could further enhance and expand upon the workshop themes.

One broad takeaway, Poghosyan stated, is that researchers have an obligation to share data responsibly to move science forward; this means they must recognize cybersecurity threats and implement appropriate data protections. Working with data takes time and resources, and scientific progress involves a solid foundation for data protection and sharing. Yann Joly, McGill University (Canada), agreed, but noted that "the devil's in the detail[s]." He noted that creating an agreed-upon international data governance framework involves work to integrate and harmonize the disparate national laws, to create incentives to motivate scientists to use responsible data sharing practices, and to design clear policies that protect researchers' data and intellectual property.

Linkov observed that each country's National Academy of Sciences operates differently. In the United States, it is an independent entity that advises the government and receives federal funding for some activities but is nongovernmental and independent. Two participants from Kyrgyzstan, Damira Ashiralieva and Jailobek Orozov, explained that the National Academy of Sciences Kyrgyzstan works directly within the Department of Education and Science. Its institutes cover a wide range of life science topics and participate in government, nongovernment, and international collaborations, always working closely with the appropriate government agencies, including lawyers from the Ministry of Justice, to ensure legal compliance. Azizakhon Haidarova of the Tajikistan Branch Office of the International Science and Technology Center (ISTC) stated that the Tajikistan National Academy of Sciences is a state organization, and that obtaining the necessary government permission to implement certain projects, especially within the framework of international exchanges and cooperation, may not always be an easy process. Haidarova noted that while she understands that obtaining proper permission is necessary for scientists in her country, there are occasions when the time frame to obtain such permission is delayed, and this can greatly complicate the work of scientists. Linkov wondered if this hierarchical nature could pose a barrier to research, and Haidarova noted that persistence and persuasion from project leaders is necessary to move science forward.

Ashiralieva expressed her appreciation for the opportunity to compare the data sharing policies of Kyrgyzstan with those of other countries, meet new potential collaborators, and learn about cybersecurity threats and practices. Haidarova and Samariddin Barotov, Tajikistan Node of the Global Biodiversity Information Facility, agreed, and both suggested that these topics could be brought to wider audiences across Central Asia by holding face-to-face meetings of legislators, lawyers, and information technology professionals, in addition to scientists and researchers, to learn from international experts and collaboratively explore data governance problems and legislative and regulatory solutions. "Such meetings should be held for a wider audience," Haidarova emphasized. "I am, of course, a patriot of my own country, and I am always concerned about its interests; but [being a] patriot does not mean talking about our country's superiority or asserting that it is problem free. A patriot will offer ideas for solving existing problems," she continued. Haidarova and Barotov noted that Tajikistan does not have a single regulation or law relating to medical data protection, and it would benefit greatly from such meetings to spread awareness of these issues among the legal community and decision-makers, in addition to scientists.

Participants offered several observations about which audiences were not present at the workshops and what other groups might be well positioned to contribute to and benefit from discussions of these issues. Linkov suggested that an inclusive approach is needed so that people of all age groups and experience levels—from students to lawmakers to subject matter experts to representatives from ministries, community organizations, and marginalized groups—can be better equipped to influence data sharing and data governance policy and practice in Central Asian countries. Ashiralieva agreed, noting that ministry representatives and institutional leaders need this information to develop research recommendations and approve curriculum changes, and students need this information to become competent in research design and data security and analysis. Namazbek Abdykerimov, Institute of Biotechnology of the National Academy of Sciences (Kyrgyzstan), suggested using a neutral and common forum, such as the ISTC, for developing additional levels of cooperation. Nurbolot Usenbaev, Republic-Level Center for Quarantine and Highly Dangerous Infections of the Ministry of Health of Kyrgyzstan, noted that the leaders of Central Asia's regional scientific groups, which design collaborative research projects that require life science data sharing, should also be included in future discussions to inform their efforts to establish guidelines for transparency and responsible data exchange.

Rita Guenther, U.S. National Academies of Sciences, Engineering, and Medicine suggested that meeting in person could enable more voices at the table to collaborate on solutions that balance benefits and risks, and deepen the conversations around these collective, continuous challenges. In-person meetings can also create more sustained cooperation and engagement, although this involves continuous time and resource demands. "I feel that it is very important to try to continue these types of discussions over time," Guenther said. "These challenges that we are facing collectively are problems and challenges that will always be with us when we speak about the balance between positive contributions to society from science, and the need to also make sure that those . . . are the only contributions and that there are no negative ramifications of that, unintended or otherwise." Linkov agreed, calling for "sustainable development, sustainable implementation, and sustainable collaboration."

EXAMPLES OF POTENTIAL NEXT STEPS

Looking forward, Linkov highlighted four ways that the best practices described during the workshop series could be implemented and advanced. First, regional or national scientific networks consisting of data repositories, scientific societies, journals, and conferences can be created. Some of these already exist or are being built, and enhancing or expanding them will be a great step forward, she said. Second, funding sources can be investigated for in-person meetings and sustained engagement, initiatives, and collaborations. Third, Central Asian scientists could consider publishing their work in journals and use repositories that are available to a wider audience internationally. For their part, Western researchers can offer scientific communication training to help overcome language or cultural barriers. Finally, Linkov said that engaging diverse groups of scientists, students, decision-makers, ministry officials, and other stakeholders in creating a more supportive environment for scientific collaboration can help to expand the reach and benefits of research.

Participants from Central Asia described several additional opportunities for enhancing responsible data practices and informing data governance more broadly. Kalysbek Kydyshov, Republic-Level Center for Quarantine and Highly Dangerous Infections of the Ministry of Health (Kyrgyzstan), suggested that students in Kyrgyzstan would benefit from an organization encouraging international research collaborations. Orozov shared that his country has a successful 6-month, all-expenses-paid internship program with the Japanese International Cooperation Agency, in which early-career scientists receive training in research, writing, and publishing.

Kydyshov also noted that a lack of data scientists in Kyrgyzstan means that while data are collected, data analysis is more limited. He suggested that better access to resources to teach data science techniques could help address this gap. Linkov agreed, noting that data analysis is important for identifying disease trends or outbreaks. She also added that, in her experience, funding is the key to successful collaborations and sustained engagement.

Several participants underscored the value of continuing to converse about these issues to learn from others, inform decisions, and spur action. Ashiralieva shared that Kyrgyzstan conducted in-person partnership reviews with other Central Asian nations and international organizations; in these reviews, participants realized that their shared challenges could be overcome with shared solutions. Ashiralieva found the in-person aspect of the reviews especially rewarding, noting, "Whenever you have tête-à-tête or face-to-face communication, it actually allows us to dig deeper; it allows us to identify underlying problems. It also gives us very specific, very salient information to draw parallels in our experiences."

Trisha Tucholski, U.S. National Academies, shared that four additional events are being planned to further disseminate the outcomes of the workshop series to decision-makers in Central Asia, and Guenther asked participants to suggest ideas for approaches and areas of focus for these meetings. Ashiralieva suggested that individuals or groups nominate subjects (e.g., cybersecurity) to cover in greater depth, rank the results, and choose the items that seem to resonate with the greatest number of people. She also suggested that more lawmakers, security experts, and other decision-makers for digitization, health, and science should be invited, as their work is relevant to these topics.

Barotov and Kydyshov suggested that meeting in person could help to improve engagement and participation from parties in Central Asia. While the next four meetings are slated to be virtual, Tucholski noted that a third phase of activity could include in-person meetings as early as 2024. Haidarova suggested that it may be easier and more cost effective for American participants to travel to Central Asian countries and offered her country as a venue for offline events, since in her opinion, it is the live format of communication that will provide more opportunities for scientists to learn and have their questions addressed. Ashiralieva suggested that such a meeting could be combined with the Biosafety Association for Central Asia and the Caucasus[1] international conference for biosecurity, which will be attended by scientists from Central Asia, Mongolia, and the Caucasus. Closing the workshop, Tucholski expressed her gratitude to the many speakers

[1] See https://internationalbiosafety.org/ifba_members/biosafety-association-for-the-central-asia-and-caucasus/.

and participants who brought their experiences and perspectives to create a fruitful exchange over the course of the events.

References

Askarova, S., B. Umbayev, A-R. Masoud, A. Kaiyrlykyzy, Y. Safarova, A. Tsoy, F. Olzhayev, and A. Kushugulova. 2020. The links between the gut microbiome, aging, modern lifestyle and Alzheimer's disease. *Frontiers in Cellular and Infection Microbiology* 10:104. https://doi.org/10.3389/fcimb.2020.00104.

Barotov, S. 2023. *The Herbarium Fund of the Institute of Botany, Plant Physiology and Genetics at the Tajikistan National Academy of Sciences—BRAHMS Records*. Last modified December 30, 2023. https://doi.org/10.15468/ntdjg9 (accessed August 4, 2023).

Boscarino, N., R. Cartwright, K. Fox, and K. S. Tsosie. 2022. Federated learning and Indigenous genomic data sovereignty. *Nature Machine Intelligence* 4(11): 909-911. https://doi.org/10.1038/s42256-022-00551-y.

Carroll, S., Garba, I., Figueroa-Rodríguez, O, Holbrook, J., Lovett, R., Materechera, S., Parsons, M., Raseroka, K., Rodriguez-Lonebear, D., Rowe, R., Sara, R., Walker, J., Anderson, J. and Hudson, M., 2020. "The CARE Principles for Indigenous Data Governance." *Data Science Journal* 19(1):43 https://doi.org/10.5334/dsj-2020-043.

Check Point Software Technologies, Ltd. 2022. Check Point Research: Cyber attacks increased 50% year over year. *Check Point Blog*. January 10, 2022. https://blog.checkpoint.com/security/check-point-research-cyber-attacks-increased-50-year-over-year (accessed January 22, 2024).

Chhay, A. 2022. New type of social engineered phishing. *Information Technology*. Lawrence Berkeley National Laboratory. https://it.lbl.gov/new-type-of-social-engineered-phishing (accessed August 4, 2023).

CIOMS (Council for International Organizations of Medical Sciences). 2016. *International ethical guidelines for health-related research involving humans*. https://doi.org/10.56759/rgxl7405.

Convention on Biological Diversity. n.d.a. *The Cartagena Protocol on Biosafety*. https://bch.cbd.int/protocol/. (accessed July 31, 2023).

Convention on Biological Diversity. n.d.b. *The Nagoya Protocol on Access and Benefit-sharing*. https://www.cbd.int/abs (accessed July 31, 2023).

Das, S., and M. Lowe. 2018. Nature read in Black and White: Decolonial approaches to interpreting natural history collections. *Journal of Natural Science Collections* 6:4-14. http://www.natsca.org/article/2509 (accessed January 22, 2024).

Davis, J. n.d. *Are natural history museums inherently racist?* Natural History Museum. https://www.nhm.ac.uk/discover/news/2019/july/are-natural-history-museums-inherently-racist.html (accessed August 4, 2023).

Dinha, F. 2023. The human factor in cybersecurity: Understanding social engineering. *Forbes*. April 10, 2023. https://www.forbes.com/sites/forbestechcouncil/2023/04/10/the-human-factor-in-cybersecurity-understanding-social-engineering/?sh=2054f5b86a02 (accessed August 4, 2023).

Eke, D. O., A. Bernard, J. G. Bjaalie, R. Chavarriaga, T. Hanakawa, A. J. Hannan, S. L. Hill, M. Martone, A. McMahon, O. Ruebel, S. Crook, E. Thiels, and F. Pestilli. 2022. International data governance for neuroscience. *Neuron* 110(4):600-612. https://doi.org/10.1016/j.neuron.2021.11.017.

EMA (European Medicines Agency). 2021. *Cyberattack on EMA—Update 5*. January 15, 2021. https://www.ema.europa.eu/en/news/cyberattack-ema-update-5 (accessed January 22, 2024).

Errington, T. M., A. Denis, N. Perfito, E. Iorns, and B. A. Nosek. 2021. Challenges for Assessing replicability in preclinical cancer biology. *ELife* 10(December):e67995. https://doi.org/10.7554/eLife.67995.

EUCAST (European Society of Clinical Microbiology and Infectious Diseases). n.d. *The European Committee on Antimicrobial Susceptibility Testing—EUCAST*. https://www.eucast.org (accessed August 3, 2023).

European Commission. n.d. *European health data space*. https://health.ec.europa.eu/ehealth-digital-health-and-care/european-health-data-space_en (accessed May 12, 2023).

Everett, D. 2023. *Data governance vs data management: What's the difference?* Informatica. Last published June 29, 2023. https://www.informatica.com/blogs/data-governance-vs-data-management-whats-the-difference.html (accessed August 4, 2023).

Funk, C., M. Hefferon, B. Kennedy, and C. Johnson. 2019. Trust and mistrust in Americans: Views of scientific experts. *Pew Research Center Science & Society Blog*. August 2, 2019. https://www.pewresearch.org/science/2019/08/02/trust-and-mistrust-in-americans-views-of-scientific-experts (accessed January 22, 2024).

GA4GH (Global Alliance for Genomics and Health). n.d. https://www.ga4gh.org (accessed August 3, 2023).

Gabelica, M., R. Bojčić, and L. Puljak. 2022. Many Researchers were not compliant with their published data sharing statement: A mixed-methods study. *Journal of Clinical Epidemiology* 150(October):33-41. https://doi.org/10.1016/j.jclinepi.2022.05.019.

GBIF (Global Biodiversity Information Facility). n. n.d.a. *Global Biodiversity Information Facility*. https://www.gbif.org (accessed August 4, 2023).

GBIF. n.d.b. Kazakhstan. https://www.gbif.org/country/KZ/summary (accessed August 3, 2023).

GO FAIR. n.d. *FAIR principles*. https://www.go-fair.org/fair-principles (accessed August 3, 2023).

India Stack. n.d. *FAQ*. https://indiastack.org/faq.html (accessed August 3, 2023).

Kaiyrlykyzy, A., S. Kozhakhmetov, D. Babenko, G. Zholdasbekova, D. Alzhanova, F. Olzhayev, A. Baibulatova, A. R. Kushugulova, and S. Askarova. 2022a. Study of gut microbiota alterations in Alzheimer's dementia patients from Kazakhstan. *Scientific Reports* 12(1):15115. https://doi.org/10.1038/s41598-022-19393-0.

Kaiyrlykyzy, A., B. Umbayev, A-R. Masoud, A. Baibulatova, A. Tsoy, F. Olzhayev, D. Alzhanova, G. Zholdasbekova, K. Davletov, A. Akilzhanova, and S. Askarova. 2022b. Circulating adiponectin levels, expression of adiponectin receptors, and methylation of adiponectin gene promoter in relation to Alzheimer's disease. *BMC Medical Genomics* 15(1):262. https://doi.org/10.1186/s12920-022-01420-8.

Local Contexts. n.d. *Grounding Indigenous rights*. https://localcontexts.org (accessed August 4, 2023).

Mackey, T. K., A. J. Calac, B. S. Chenna Keshava, J. Yracheta, K. S. Tsosie, and K. Fox. 2022. Establishing a blockchain-enabled Indigenous data sovereignty framework for genomic data. *Cell* 185(15):2626-2631. https://doi.org/10.1016/j.cell.2022.06.030.

McCartney, A. M., M. A. Head, K. S. Tsosie, B. Sterner, J. R. Glass, S. Paez, J. Geary, and M. Hudson. 2023. Indigenous peoples and local communities as partners in the sequencing of global eukaryotic biodiversity. *NPJ Biodiversity* 2(1):1-12. https://doi.org/10.1038/s44185-023-00013-7.

MDL. 2017. NotPetya Ransomware disrupts Merck vaccine production. *Cybersecurity*. University of Hawai'i–West O'ahu. https://westoahu.hawaii.edu/cyber/regional/gce-us-news/notpetya-ransomware-disrupts-merck-vaccine-production (accessed August 4, 2023).

NASEM and NAM (National Academies of Sciences, Engineering, and Medicine and National Academy of Medicine). 2023. *Toward equitable innovation in health and medicine: A framework*. Washington, DC: The National Academies Press. https://doi.org/10.17226/27184.

National Institute of Standards and Technology. n.d. *Cybersecurity*. Computer Security Resource Center. U.S. Department of Commerce. https://csrc.nist.gov/glossary/term/cybersecurity (accessed August 4, 2023).

Native BioData Consortium. n.d. https://nativebio.org (accessed August 4, 2023).

NIH (National Institutes of Health). n.d. *2023 NIH Data Management and Sharing Policy*. https://oir.nih.gov/sourcebook/intramural-program-oversight/intramural-data-sharing/2023-nih-data-management-sharing-policy (accessed August 3, 2023).

NITI Aayog (National Institution for Transforming India). 2020. *Data empowerment and protection architecture: Draft for discussion*. https://www.niti.gov.in/sites/default/files/2020-09/DEPA-Book.pdf (accessed January 22, 2024).

Osborne, C. 2021. *Oxford University lab with COVID-19 research links targeted by hackers*. ZDNET. https://www.zdnet.com/article/oxford-university-biochemical-lab-involved-in-covid-19-research-targeted-by-hackers (accessed August 4, 2023).

Pain, E. 2016. French institute agrees to share patent benefits after biopiracy accusations. *Science*. February 10, 2016. https://www.science.org/content/article/french-institute-agrees-share-patent-benefits-after-biopiracy-accusations (accessed August 4, 2023).

PHA4GE (Public Health Alliance for Genomic Epidemiology). n.d. https://pha4ge.org (accessed July 2, 2021).

Puttaswamy, J. K. S. n.d. In the Supreme Court of India Civil Original Jurisdiction. https://www.humandignitytrust.org/wp-content/uploads/resources/Puttaswamy-v.-Union-of-India-full-judgment.pdf (accessed on February 13, 2024).

Re3data.org (Registry of Research Data Repositories). n.d. *GISAID*. https://doi.org/10.17616/R3Q59F (accessed January 22, 2024).

Ryabushkina, N. A., S. Abugalieva, and Y. Turuspekov. 2016. [Problems of study and conservation of flora biodiversity in Kazakhstan] ПРОБЛЕМА ИЗУЧЕНИЯ И СОХРАНЕНИЯ БИОРАЗНООБРАЗИЯ ФЛОРЫ КАЗАХСТАНА. *Eurasian Journal of Biotechnology* 3(July):13-23. https://doi.org/10.11134/btp.3.2016.2.

Tedersoo, L., R. Küngas, E. Oras, K. Köster, H. Eenmaa, A. Leijen, M. Pedaste, M. Raju, A. Astapova, h. Lukner, K. Kogermann, and T. Sepp. 2021. Data sharing practices and data availability upon request differ across scientific disciplines. *Scientific Data* 8(1):192. https://doi.org/10.1038/s41597-021-00981-0.

UNESCO (United Nations Educational, Scientific and Cultural Organization). 2017. *Recommendation on Science and Scientific Researchers*. https://www.unesco.org/en/legal-affairs/recommendation-science-and-scientific-researchers (accessed August 3, 2023).

UNESCO. 2023. *UNESCO recommendation on open science.* https://www.unesco.org/en/open-science/about (accessed August 3, 2023).

Vos, T., S. S. Lim, C. Abbafati, K. M. Abbas, M. Abbasi, and the GBD 2019 Diseases and Injuries Collaborators. 2020. Global burden of 369 diseases and injuries in 204 countries and territories, 1990–2019: A systematic analysis for the Global Burden of Disease Study 2019. *The Lancet* 396(10258):1204-1222. https://doi.org/10.1016/S0140-6736(20)30925-9.

WHO (World Health Organization). 2010. *Responsible life sciences research for global health security: A guidance document.* https://www.ncbi.nlm.nih.gov/books/NBK305040 (accessed January 22, 2024).

WHO. 2022a. *Global guidance framework for the responsible use of the life sciences: Mitigating biorisks and governing dual-use research.* https://www.who.int/publications-detail-redirect/9789240056107 (accessed August 4, 2023).

WHO. 2022b. *Sharing and reuse of health-related data for research purposes: WHO policy and implementation guidance.* https://www.who.int/publications-detail-redirect/9789240044968 (accessed August 3, 2023).

Zhunussova, G., G. Afonin, S. Abdikerim, A. Jumanov, A. Perfilyeva, D. Kaidarova, and L. Djansugurova. 2019. Mutation spectrum of cancer-associated genes in patients with early onset of colorectal cancer. *Frontiers in Oncology* 9(August):673. https://doi.org/10.3389/fonc.2019.00673.

Appendix A
Statement of Task

The National Academies of Sciences, Engineering, and Medicine (NASEM) will convene an ad hoc planning committee of international scientists to conduct a series of workshops to identify best practices for protecting biological and life science data and knowledge while promoting the sharing of these scientific data. The committee will engage early-career and established scientists in the project by inviting them to prepare presentations on the following topics:

- Risks and vulnerabilities affecting the protection of biological and life science data and knowledge;
- Existing international and national policies, practices, and norms related to access and transfer of biological and life science data, focusing on sharing of data, protecting data, and current associated challenges;
- Existing and needed practices for cyber, data, and information security and evaluation of data inquiries; and
- Effectiveness and ability to translate the practices in different institutional contexts.

Invited scientists will be given opportunities to present their work to U.S. and international biorisk experts and scientists, and the committee. Based primarily on the invited presentations and discussions and supplemented with a review of the literature, as appropriate, NASEM will produce a Proceedings of the workshop discussions.

Appendix B
Workshop Agendas

ENGAGING SCIENTISTS IN CENTRAL ASIA ON LIFE SCIENCES DATA GOVERNANCE PRINCIPLES—A WORKSHOP SERIES

Increasing digitalization in the life sciences has led to numerous opportunities for addressing challenges in health, agricultural, environmental, and other sectors. Life scientists are generating and analyzing large amounts of data from various sources, and although data from scientific research are not necessarily sensitive, they may represent the outcomes of significant investment at individual, institutional, governmental, and industry levels. To protect these investments while also enabling scientific advancement, the U.S. National Academies of Sciences, Engineering, and Medicine is conducting a six-part virtual workshop series to better understand effective practices for the management and protection of life sciences data and knowledge within research institutions. This virtual workshop series aims to engage early-career and established scientists from Central Asia, the United States, and elsewhere to exchange experiences and collaboratively identify best practices for protecting biological and life science data and knowledge, while promoting the responsible sharing of these scientific data. Each consecutive workshop session builds on the previous session, and participants are encouraged to attend as many sessions as possible to ensure continuity of discussion and to gain maximum benefit.

MAY 4, 2023

10:00 am EST — Welcome and Opening Remarks
Trisha Tucholski, U.S. National Academies Board on Life Sciences (Project Lead)
Lusine Poghosyan, Columbia University (Committee Chair)

10:05 — Introduction
Vasiliki Rahimzadeh, Baylor College of Medicine (Committee Member)
Life Science Data Governance: Principles, Policies, and Practices

10:15 — Roundtable: Life Science Research in Central Asia
Moderator: Damira Ashiralieva, Ministry of Health of Kyrgyzstan (Committee Member)

Goals:

- Highlight life science and biological research efforts across Central Asia
- Explore the scale and scope of life science and biological data in Central Asia

Speakers:
Jailobek Orozov (Committee Member)
Director, A. Duisheev Kyrgyz Research Institute of Veterinary Science
Bishkek, Kyrgyzstan
Determination of Bovine Brucellosis Genotypes in Kyrgyzstan

Margarita Ishmuratova (Committee Member)
Professor of Botany, Head of the Research Park of Biotechnology and Eco-monitoring
Buketov Karaganda University
Karaganda, Kazakhstan
The Study of Flora and Fauna in Kazakhstan

Ravshan Azimov
Senior Lecturer, Center for Retraining and Professional Development of Teaching Staff at the Tashkent Medical Academy
Ministry of Higher and Secondary Special Education of Uzbekistan
Tashkent, Uzbekistan
Data Governance for Life Science in Uzbekistan

Nurbolot Usenbaev
Deputy Director, Republic-Level Center for Quarantine and Highly Dangerous Infections
Ministry of Health of Kyrgyzstan
Bishkek, Kyrgyzstan
Microbiotes in Central Asia and Caucasus

Gulnur Zhunussova
General Director, Institute of Genetics and Physiology
Almaty, Kazakhstan
Cancer Genetics and Genomics Studies in Kazakhstan

Sholpan Askarova (Committee Member)
Head of the Laboratory of Bioengineering and Regenerative Medicine
Nazarbayev University
Almaty, Kazakhstan
Alzheimer's Disease Research in Kazakhstan

Faina Linkov (Committee Member)
Professor of Health Administration and Public Health
Duquesne University
Pittsburgh, PA, United States
General Research Challenges in Central Asia

11:25 Guided Discussion: Data Practices and Norms for Life Science in Central Asia
Moderator: Lusine Poghosyan

Panelists: Jailobek Orozov, Margarita Ishmuratova, Ravshan Azimov, Nurbolot Usenbaev, Gulnur Zhunussova, Sholpan Askarova, Faina Linkov

Goals:

- Understand the context-specific national and institutional data practices and norms for life science research in Central Asia
- Identify existing and emerging challenges related to data management for life sciences research in the region
- Explore emerging data governance tools and trends in Central Asia

Discussion Questions:

- Are the data practices (collection, storage, processing, curation, sharing, access, and use) discussed by individual researchers representative of national or institutional practices? Do practices differ between scientific disciplines or sectors?
- Can you describe some of the challenges you have experienced in your research and how they contribute to your data practices? Are the challenges unique to Central Asia or common internationally?
- To what degree do issues exist with authorized and unauthorized data access? What policies and practices are in place for enabling data sharing while preventing unauthorized access?
- Are there cases when scientific research or data is not covered under specific governing policies?
- What are some emerging methodology, platforms, and technologies for managing and governing data in Central Asia?

11:55 Summary of Key Takeaways
Trisha Tucholski

12:00 pm Closing Remarks
Lusine Poghosyan

MAY 11, 2023

10:00 am EST Welcome and Opening Remarks
Trisha Tucholski, U.S. National Academies Board on Life Sciences (Project Lead)
Lusine Poghosyan, Columbia University (Committee Chair)

10:05 Introduction
Yann Joly, McGill University (Committee Member)
Data Governance Principles for Life Science Across the Globe

10:20 Roundtable: Data Governance Experts from Across the Globe
Moderator: John Ure, Access Partnership (Committee Member)

Goals:

- Highlight data governance laws, tools, and frameworks used across the globe
- Explore norms and emerging trends in data governance approaches for life science and biological research in different areas of the world

Speakers:
Taunton Paine
Director, Scientific Data Sharing Policy Division, Office of Science Policy
National Institutes of Health (NIH)
Bethesda, MD, United States
Federal and NIH Policies for Managing and Sharing Research Data

Fruszina Molnár-Gábor
Professor, International and Health Law and Data Protection Law
Heidelberg University
Heidelberg, Germany
General Data Protection Regulation with Respect to Life Science and Biological Data Sharing

Athira P.S.
Director, Centre for Intellectual Property Rights
National University of Advanced Legal Studies
Kerala, India
Life Science and Medical Research in India: Relevance of Robust Data Governance Principles and Structures

Hellen Nansumba
Biorepository Manager, National Health Laboratory Services and Diagnostics
Ministry of Health of Uganda
Kampala, Uganda
Overview of the Role of Regulatory and Governance Frameworks in Human Biological Materials and Data Sharing in the National Biorepository in Uganda

Ya-Hsin Li
Associate Professor and Director, Department of Health Policy and Management
Office of International Affairs
Chung-Shan Medical University, Taiwan
Taiwan's National Health Insurance Research Database

11:20	Guided Discussion: Compare and Contrast Approaches to Data Governance Moderator: Wei Zheng, Vanderbilt University (Committee Member) Panelists: Taunton Paine, Fruszina Molnár-Gábor, Athira P.S., Hellen Nansumba, Ya-Hsin Li Goals: • Identify similarities and differences in different approaches to data governance presented by the panel • Identify gaps and challenges in existing data governance policies and practices Discussion Questions: • What general points of agreement can we identify between the policies and practices for various countries? • Taking a broad comparative view on the biological data sharing framework, what differences between featured countries, if any, appear particularly challenging to reconcile? • Are there some outliers (countries that stand out because of a particularly distinct data governance regime)? How can we explain their situation?
11:55	Summary of Key Takeaways Trisha Tucholski
12:00 pm	Closing Remarks Lusine Poghosyan

JUNE 1, 2023

10:00 am EST	Welcome and Opening Remarks Trisha Tucholski, U.S. National Academies Board on Life Sciences (Project Lead) Lusine Poghosyan, Columbia University (Committee Chair)
10:10	Introduction Krystal Tsosie Assistant Professor, Arizona State University (Committee Member) *Indigenous Knowledges, Biological Stewardship, and Community Data Governance*
10:30	Zhyldyz Tegizbekova Associate Professor of Law, Ala-Too International University, Kyrgyzstan

Challenges of Traditional Knowledge Preservation and Protection in Central Asia

Questions for consideration:

- Is the intellectual property system sufficient to protect traditional knowledge in Central Asian countries?
- Are special legal measures needed for safeguarding traditional knowledge?
- What is the role of states to preserve and protect traditional knowledge?

10:50 Samariddin Barotov
Institute of Botany, Plant Physiology, and Genetics
Head, Global Biodiversity Information Facility (GBIF), Tajikistan
GBIF – Kick-starting the Biodiversity Publication Process for Tajikistan

11:00 Guided Discussion
Moderator: Vasiliki Rahimzadeh, Baylor College of Medicine (Committee Member)

Goals:

- Consider the benefits together with the risks and vulnerabilities of sharing life science and biological data
- Discuss best practices for sharing data feasibly and equitably
- Understand how to navigate policies for authorized sharing/access versus unauthorized access

Panelists:
Krystal Tsosie (Committee Member), Arizona State University, United States
Zhyldyz Tegizbekova, Ala-Too International University, Kyrgyzstan
Samariddin Barotov, Global Biodiversity Information Facility, Tajikistan
Margarita Ishmuratova (Committee Member), Buketov Karaganda University, Kazakhstan

Discussion Questions:

- What are some of the benefits of making data openly available and accessible?
- What are some potential risks and/or drawbacks associated with open access data? With sharing data? Who incurs these risks?
- How can risks be managed such that benefits from sharing research are maximized?
- What protection mechanisms are currently in place to protect data that should be protected, while also ensuring that it can be appropriately utilized and/or shared?

- What practices exist to help scientists understand how to prevent unauthorized access to knowledge and data?
- In addition to scientists and researchers, who is in a position to allow proper access and prevent improper access (e.g., computer/IT personnel, university administration, public health officials)?
- Who is in a position to make changes or improve current policies of practices for data protection?

11:50 Summary of Key Takeaways
Trisha Tucholski

12:00 pm Closing Remarks
Lusine Poghosyan

JUNE 8, 2023

10:00 am EST Welcome and Opening Remarks
Trisha Tucholski, U.S. National Academies Board on Life Sciences (Project Lead)

10:15 Roundtable: Life Science Data Governance in Central Asia
Moderator: Trisha Tucholski, U.S. National Academies Board on Life Sciences

Goals:

- Review existing national policies, practices, and norms related to sharing and protection of biological data in Central Asia
- Identify the associated challenges for sharing and protecting data given the existing policies, practices, and norms

Speakers:
Damira Ashiralieva (Committee Member)
Virologist, National Scientific-Practical Center
Ministry of Health of the Kyrgyzstan
Bishkek, Kyrgyzstan
Ensuring Information Security in Kyrgyzstan

Shakhlo Turdikulova
Deputy Minister, Ministry of Higher Education, Science, and Innovation
Director, Center for Advanced Technologies
Tashkent, Uzbekistan
Biological Data Sharing in Uzbekistan
Zulfiya Davlyatnazarova
Deputy Director, Institute of Botany, Plant Physiology & Genetics
Tajikistan National Academy of Sciences

Dushanbe, Tajikistan
Biological Data Sharing in Tajikistan

11:00 Panel of Experts: Life Science Data Governance in Practice
Moderator: Rita Guenther, U.S. National Academies, Policy and Global Affairs

Goals:

- Understand how existing policies, practices, and norms translate to real-life scenarios in the field, clinic, and laboratory
- Explore how existing data governance policies, practices, and norms could be improved or standardized to maximize appropriate sharing of the data with a range of stakeholders

Panelists:
Pavel Tarlykov
Head, Laboratory of Proteomics and Mass Spectrometry
National Center for Biotechnology
Astana, Kazakhstan

Eastwood Leung
Adjunct Professor, International University of Central Asia
Tokmok, Kyrgyzstan

Elmira Utegenova
Deputy Director, Laboratory of Diagnostic Service
Almaty, Kazakhstan

Discussion Questions:

- What are the types of data that you currently work with? Are there any unique features associated with the data or limitations on its use?
- Who has access to these data (in the field, in the lab, within partnerships, the government, for-profits, others)? Who determines this? Why?
- What are the current policies and practices that apply to the collection, processing, analysis, storage, sharing and protection of your data?
- Are there any cases where existing policies, practices, or norms do not necessarily cover specific data types or cases? Are there gaps in current policies, practices, or norms that are needed to better obtain/maintain the balance of the responsible/appropriate sharing of data with the relevant protection of data?
- To what degree do issues exist with authorized and unauthorized data access? How is this determined? What policies, practices, or

norms are in place for enabling data sharing and preventing unauthorized access?
- What would need to happen to improve or standardize current data governance policies, practices, or norms? Who is in a position to make those changes? Why?

11:50 Summary of Key Takeaways
Trisha Tucholski, U.S. National Academies Board on Life Sciences (Project Lead)

12:00 pm Closing Remarks
Lusine Poghosyan, Columbia University (Committee Chair)

JUNE 15, 2023

10:00 am EST Welcome and Opening Remarks
Trisha Tucholski, U.S. National Academies Board on Life Sciences (Project Lead)
Lusine Poghosyan, Columbia University (Committee Chair)

10:10 Introduction
Gautham Venugopalan
Director of International Engagement, Gryphon Scientific (Committee Member)
Cyber Risk Management in Life Sciences Research

10:30 Case Study Discussion
Trisha Tucholski

10:45 Data and Information Security in Central Asia
Moderator: Kavita Berger, U.S. National Academies Board on Life Sciences (Director)

Goals:
- Discuss existing security practices and their associated challenges
- Identify needed practices for cyber, data, and information security
- Discuss differences in risks and practices among institution types and fields.

Discussion Questions:
- What are some issues to consider in securing life science and biological data and associated innovations?
- What are existing security practices that may reduce the risk of exploitation or unauthorized access to life science and biological data and knowledge? Are there identifiable gaps or challenges in current practices?

- What would researchers or IT staff need (training, information, tools, applications, risk assessments, people, etc.) to be able to implement security practices?
- Are the security practices discussed today translatable in different national or institutional contexts? If not, why not?
- Who is in a position to improve cyber- and information security measures or practices?

11:45 Summary of Key Takeaways
Trisha Tucholski

12:00 pm Closing Remarks
Lusine Poghosyan

JUNE 20, 2023

11:00 am EST Welcome and Opening Remarks
Trisha Tucholski, U.S. National Academies Board on Life Sciences (Project Lead)

11:05 Summary of the Workshop Series
Lusine Poghosyan, Columbia University (Committee Chair)

11:15 What Have We Learned? A Review
The Workshop Planning Committee

11:30 What's Next?
Moderator: Faina Linkov
Associate Professor, Duquesne University (Committee Member)

12:45 pm Preview of Dissemination Events
Trisha Tucholski

1:00 Closing Remarks
Lusine Poghosyan

Appendix C
Workshop Planning Committee Biographies

Dr. Lusine Poghosyan is Stone Foundation and Elise D. Fish professor of nursing and executive director of the Center for Healthcare Delivery Research & Innovations at Columbia University School of Nursing and professor of health policy and management at Columbia University Mailman School of Public Health. She is an internationally recognized nurse scientist with health services research expertise. She is principal investigator on multiple federal- and foundation-funded (e.g., National Institutes of Health, Agency for Healthcare Research and Quality, Robert Wood Johnson Foundation) research projects, focused on primary care, health care workforce, and quality of care. She utilizes large databases containing records on millions of patients and clinicians. Dr. Poghosyan is a fellow of the American Academy of Nursing and the New York Academy of Medicine. She was past chair of the Primary Care Expert Panel at the American Academy of Nursing and AcademyHealth's Interdisciplinary Research Group on Nursing Issues Advisory Board. She is a recipient of multiple awards, including the International Nurse Researcher Hall of Fame from Sigma Theta Tau International in 2022. Dr. Poghosyan received her PhD from the University of Pennsylvania, MPH from the American University of Armenia, and BSN from Erebuni Medical College.

Dr. Damira Omurzakovna Ashiralieva is a virologist at the National Scientific-Practical Center of Infection Control ("Preventative Medicine"), a scientific production organization under the Ministry of Health of Kyrgyzstan. Dr. Ashiralieva previously conducted research on the use of polymerase chain reaction to identify infectious diseases and disease resistance in Moscow, Russia. Dr. Ashiralieva's current responsibilities include detecting problems in the laboratory diagnosis of infectious diseases, including implementation of modern standards of laboratory diagnosis for healthcare-associated infections, and the development of recommendations and standard operating procedures (SOPs) in the field of laboratory diagnosis and biosafety. She established a clinical bacteriological laboratory, and in 2015, implemented requirements from the European Committee on Antimicrobial Susceptibility Testing (EUCAST) and standards of bacteriological urinalysis in Kyrgyzstan. Dr. Ashiralieva is the author of an introductory bacteriology atlas, training curriculums in antibiotic resistance, and SOPs for sample collection, storage, and transportation for new hospital units. In 2020, she received the Presidential Medal for her work combating COVID-19, and in 2016, she was named a Distinguished Physician by the Ministry of Health. In 2022, Dr. Ashiralieva was elected President of the Biosafety Association for Central Asia and the Caucasus. Dr. Ashiralieva received her doctorate in sanitation and hygiene from the Kyrgyz State Medical Institute in 1996, and her doctorate in clinical microbiology from the Institute for Medical Research, National Institutes of Health, Kuala Lumpur, Malaysia, in 2020.

Dr. Sholpan Askarova is head of the Laboratory of Bioengineering and Regenerative Medicine at National Laboratory Astana, Nazarbayev University (Astana, Kazakhstan). Her primary research interests are associated with molecular and cellular mechanisms of neurodegenerative diseases and aging, including cell signaling pathways and cytotoxic effects of amyloid-beta-peptide implicated in Alzheimer's disease (AD), genetic and epigenetic factors that may contribute to cognitive impairment and development of AD, and study of the human microbiome as a potential prognostic biomarker of AD. She is also interested in studying the molecular mechanisms of stem cell aging, rejuvenation, and development and improving the methods of cell therapy for the treatment of chronic diseases and tissue regeneration. Dr. Askarova graduated from al-Farabi Kazakh National University with a BS in 1996 and with a MS in 1999. In 2004, she received a PhD in biological sciences. From 2006 to 2011, as a "Bolashak" program scholar, Dr. Askarova joined the University of Missouri to pursue a PhD degree in biological engineering. In 2011, as a visiting researcher at the McGowan Institute for Regenerative Medicine at the University of Pittsburgh, she started her specialization in the field of regenerative medicine.

Dr. Margarita Yulayevna Ishmuratova is professor of the Botany Department and Head of the Research Park of Biotechnology and Eco-Monitoring in the Faculty of Biology and Geography at Buketov Karaganda University (2015-present). She previously served as Vice Dean for Research (2016-2020), Research Associate in the Institute of Phytochemistry (1997-2008), Leading Research Associate and Director of Zhezkazgan botanical garden (2008-2012), and Assistant Professor in the Pharmacology Department and Head of the Science Department (2012-2015). She is also editor-in-chief of the journal *Bulletin of the Karaganda University* and the *Biology*, *Medicine*, and *Geography* bulletin series, where she draws upon more than 20 years' research experience in botany and pharmacognosy, plant anatomy and morphology, plant introduction, and floristic research, and participation in state (Science Committee) and international (International Science and Technology Center) grants and programs. She is a member of the Kazakhstan Biodiversity Preservation Association and was a nominee of a state scholarship for talented young researchers in 2004, 2006, and 2008. She was awarded Best High School Teacher by the Ministry of Education and Science of Kazakhstan in 2019. Her education history includes numerous professional development courses, including: Learning and Teaching in Higher Education (Newcastle University, 2016); Mathematic Methods in Ecology and Environmental Management (Czech Technical University, 2017); Teaching Seminar of Britain Council Research Management and Administration (2016); Professional Training at the Institute of Plant Biology and Biotechnology (Almaty) on Techniques of Micro-clonal Reproduction and Cryopreservation (2020); Professional Online Training in Digital Technology in High School Institutions of the European Union; Quality of Education Using Digital Technology at the Institute of International Education (Charles University, 2020); and Professional Development on Modern Botany (Altai State University, 2020).

Dr. Yann Joly is research director of the Centre of Genomics and Policy, full professor in the Faculty of Medicine and Health Sciences in the Department of Human Genetics, and an associate member of the Bioethics Unit and the Law Faculty at McGill. He was named advocatus emeritus by the Quebec Bar in 2012 and fellow of the Canadian Academy of Health Sciences in 2017. Dr. Joly's research interests lie at the interface of the fields of data sciences, human rights law, and bioethics. He is the current co-chair of the Regulatory and Ethics Work Stream of the Global Alliance for Genomics and Health. He created the world's first international Genetic Discrimination Observatory (see: https://gdo.global/en/gdo-description) in 2018. The observatory is represented in 25 countries in five world regions, including China, Kazakhstan, and Ukraine. Dr. Joly has published his findings in more than 200 peer-reviewed articles featured in top legal, ethical and scientific journals. He served as a legal advisor on multiple institutional review boards in the public and private sectors. In 2012, Dr. Joly received the Quebec Bar Award of Merit for his work on data sharing and privacy.

Dr. Faina Linkov is associate professor and chair at Duquesne University, Pittsburgh, Pennsylvania. She is a multidisciplinary researcher with research interests in molecular epidemiology, cancer, prevention, health systems research, global health, scientific communications, and research productivity. She has been active in initiating research work with several groups in Asia, publishing on the public health challenges in Kazakhstan. Dr. Linkov's work helped to spread the word about the need for endometrial cancer prevention in Central Asia and specifically in Kazakhstan. Pipelle biopsy, a commonly used diagnostic modality in the United States for endometrial cancer detection and control, is rarely practiced in Kazakhstan. Her protocol for investigating various factors impacting the use of pipelle biopsy were shared with cancer researchers at Nazarbayev University in Kazakhstan and became the focus on several recent publications. Dr. Linkov's research career has been extremely productive, with more than 100 original research publications and reviews. She was named fellow of the American Association for the Advancement of Science in 2020, the first female faculty member to have such distinction at Duquesne University. Her H-index in Google Scholar is 25, indicating a high degree of citation of her work. Dr. Linkov obtained her PhD in epidemiology at the University of Pittsburgh, where she conducted research from 2005 to 2020.

Dr. Jailobek Chokonovich Orozov currently serves as director at A. Duisheev Kyrgyzstan Research Institute of Veterinary Sciences. In 1997, he began his career at the Department of Veterinary Medicine clinic of the Kyrgyzstani Agrarian Academy. In 2001, he worked at the Kyrgyzstan Research Institute of Animal Husbandry, Veterinary Sciences, and Pastures in the Laboratory of Virology and Biotechnology as a junior researcher. In 2009, he worked as a senior researcher at the Laboratory of Virology and Biotechnology of the A. Duisheev Kyrgyzstan Research Institute of Veterinary Sciences, serving as its deputy director from 2017-2019. In 2011, he managed a project entitled *Epizootological bio-monitoring of sheep and goat pox in Kyrgyzstan* (International Science and Technology Center KR-1867). In 2014, he worked as a lead specialist at the International Educational Center of KNAU. From 2014 to 2017, he worked as a local consultant on education and capacity building at the KNAU Department of Agricultural Projects Implementation of the Kyrgyzstan Ministry of Agriculture, Processing Industry, and Land Reclamation. He entered the K.I. Scryabin Kyrgyz National Agrarian University (KNAU) in 1992 and graduated in 1997. In 2013, he successfully defended his PhD thesis entitled *Optimization of the Polymerase Chain Reaction in the Diagnosis of Sheep and Goat Pox*, receiving his degree in biological sciences.

Dr. Vasiliki Rahimzadeh is assistant professor with the Center for Medical Ethics and Health Policy at Baylor College of Medicine. She is an active member of the Global Alliance for Genomics and Health, an international standards-setting organization dedicated to global sharing of genomic data, where she contributes to harmonization of responsible data sharing practice and policy in genomics. Dr. Rahimzadeh's research interests center on the ethical, legal, and social issues associated with sharing genomic and related health data across international borders and computing environments. Her translational research on the ethics of health data sharing has been recognized nationally and internationally with the Governor General's Gold Medal, Gordan A. Maclachlan Prize, and the Vanier Canada Graduate Scholarship. In 2019, Dr. Rahimzadeh completed a postdoctoral fellowship at the Stanford Center for Biomedical Ethics with support from an Ethical, Legal, and Social Implications program training grant from the National Human Genome Research Institute. She earned both her MS and PhD in biomedical ethics from McGill University and completed her BS in microbial biology at the University of California, Berkeley.

Dr. Rahimzadeh is an active member of the Global Alliance for Genomics and Health, a standards setting organization dedicated to global sharing of genomic data and has been involved in drafting data sharing policies.

Dr. Krystal Tsosie (Diné/Navajo Nation) is an Indigenous geneticist-bioethicist and assistant professor at Arizona State University in the School of Life Sciences. As an advocate for Indigenous genomic and data sovereignty, she cofounded the first U.S. Indigenous-led biobank, a 501(c)(3) nonprofit research institution called the Native BioData Consortium. Much of her current research centers on ethical engagement with Indigenous communities in precision health and genomic medicine. She also incorporates biostatistics, genetic epidemiology, public health, and computational approaches to health disparities and, increasingly, environmental data science and stewardship. She currently serves on the Government Policy and Advocacy Committee for the American Society of Human Genetics. Dr. Tsosie earned a MA in bioethics for studying genetic controversies in Indigenous communities and a MPH in epidemiology for studying genetic variation related to uterine fibroids. Her PhD in genomics and health disparities focused on critical ethical issues related to Indigenous data sovereignty, data governance models, and Indigenous community perceptions related to genomic data sharing.

Dr. Tsosie accepted a one-time speaker honorarium from Regeneron for serving as a guest speaker at the October 2022 DRIFT (Discovery Research Investigating Founder Population Traits) Symposium to inform the company on improving their interactions with Indigenous communities.

Dr. John Ure is an economist specializing in digital technologies and issues of data governance and sharing. He runs a data sharing research project in Hong Kong. Dr. Ure was professor at the University of East London (UK) before joining the University of Hong Kong in the 1990s, where he was also director of the Telecoms Research Project and author of two books on telecommunications in Asia, as well as many academic papers and on the editorial advisory boards of two major journals. He founded TRPC Pte Ltd in Singapore in 2007, which in 2021 became part of Access Partnership, of which he is director. Dr. Ure has more than 20 years of experience consulting for governments, United Nations (UN) agencies, regional intergovernmental organizations, and the private sector, mostly in Asia. In 2005, he helped the UN Economic and Social Commission for Asia and the Pacific (UNESCAP) create the Asia Pacific Centre for Information and Communications Technology in Incheon, South Korea, and has written modules and taught for the UN Asian and

Pacific Training Center for Information and Communication Technology in several countries. In 2021, he wrote a proposal for a Digital Solutions Centre for UNESCAP and the Government of Kazakhstan. In 2020–2021, Dr. Ure was project manager for the Intermodal Transport Data Sharing Programme in Hong Kong (Data Trust 1.0) in collaboration with the University of Hong Kong. He is currently collaborating with the Hong Kong Polytechnic University Computer Science Department in a scaled-up version (Data Trust 2.0).

Dr. Gautham Venugopalan currently serves as director of international engagement at Gryphon Scientific. His work focuses on building global capacity to address topics such as emerging technology, laboratory biosafety and biosecurity, cybersecurity, and dual-use research risks. He also has strong expertise in data science and science policy and has provided technical leadership to several emerging technology assessments and research efforts to help maximize the benefits of research while minimizing risks. Dr. Venugopalan previously served as a diplomat, working to build research collaborations between U.S. and international scientists and to develop U.S. policies on issues such as data sharing and diversity and equity in science. He has a BS in mechanical engineering from the Rose-Hulman Institute of Technology and a PhD in bioengineering from the University of California, Berkeley and University of California, San Francisco.

Dr. Wei Zheng is professor and director of Vanderbilt Epidemiology Center. He is an experienced principal investigator (PI) with a major research focus on evaluating environmental exposures, lifestyle factors, genetics, and biomarkers for risk of cancer and other chronic diseases. Dr. Zheng has been a PI or joint PI for more than 35 large National Institutes of Health–funded research grants. Over the past 15 years, he has initiated six large international epidemiologic and genetic research consortia as PI, overseeing the harmonization, generation, and/or sharing of data from more than 1.5 million study participants. Dr. Zheng has published more than 1,200 research papers and has been named as a Highly Cited Researcher by Clarivate Analytics. He has served on many professional committees and journal editorial boards. He is an elected member of the American Epidemiology Society and a recipient of the National Cancer Institute MERIT award (2009) and Vietnam Ministry of Health's Memorabilia Medal (2019), the country's highest honorable medal to foreign scientists for their contribution to public health. Dr. Zheng received his medical degree from Fudan University Shanghai Medical School and his PhD in epidemiology from the Johns Hopkins University.

Appendix D
Workshop Participants

The following lists include individuals who participated in one or more of the workshop sessions, organized by country of affiliation.

Canada

- Yann Joly (Committee Member), McGill University

Germany

- Fruszina Molnar-Gabor (Speaker), Heidelberg University

India

- Athira PS (Speaker), National University of Advanced Legal Studies

Kazakhstan

- Dina Abbasova, Scientific-Practical Centre for Sanitary-Epidemiological Expertise and Monitoring
- Shalkar Adambekov, Kazakhstan National University
- Sholpan Askarova (Committee Member), Nazarbayev University
- Batyrbek Assembekov, Asfendiarov Kazakhstan National Medical University
- Assel Bukharbayeva, Asfendiarov Kazakhstan National Medical University
- Margarita Ishmuratova (Committee Member), Karaganda Buketov University
- Alexandr Ivankov, Independent Researcher
- Ulyana Kirpicheva, Scientific-Practical Centre for Sanitary-Epidemiological Expertise and Monitoring
- Almagul Kushugulova, Nazarbayev University
- Almira Kustubayeva, Al Farabi Kazakh National University
- Zhaniya Nurgalieva (Paper Contributor), World Health Organization
- Aizada Nurkaskaeva, Scientific-Practical Centre for Sanitary-Epidemiological Expertise and Monitoring
- Mariya Prilutskaya, Semey Medical University
- Pavel Tarlykov (Panelist, Paper Contributor), National Center for Biotechnology
- Marat Tavassov (Paper Contributor), Nazarbayev University
- Nazym Tleumbetova, Scientific-Practical Centre for Sanitary-Epidemiological Expertise and Monitoring
- Zhanna Shapieva (Paper Contributor), Scientific-Practical Centre for Sanitary-Epidemiological Expertise and Monitoring
- Elmira Utegenova (Panelist), Scientific-Practical Centre for Sanitary-Epidemiological Expertise and Monitoring
- Gulnur Zhunussova (Speaker), Institute of Genetics and Physiology

Kyrgyzstan

- Namazbek Abdykerimov, Institute of Biotechnology
- Damira Ashiralieva (Committee Member), Ministry of Health of Kyrgyzstan
- Dzhainagul Baiyzbekova, National Institute of Public Health
- Gulnara Kabaeva (Paper Contributor), Information Technology Institute
- Kalysbek Kydyshov (Speaker), Ministry of Health of Kyrgyzstan
- Hon-Chui Eastwood Leung (Panelist), International University of Central Asia
- Erkin Mirrakhimov, Kyrgyzstan State Medical Academy
- Jailobek Orozov (Committee Member), A. Duisheev Kyrgyzstan Scientific Research Veterinary Institute
- Nurbolot Usenbaev (Speaker), Ministry of Health of Kyrgyzstan
- Zhyldyz Tegizbekova (Speaker), Ala-Too International University

Singapore

- John Ure (Committee Member), Access Partnership Ltd

Tajikistan

- Samariddin Barotov (Speaker, Paper Contributor), Institute of Botany, Plant Physiology, and Genetics
- Abdulaziz Davlatov, E.N. Pavlovsky Institute of Zoology and Parasitology
- Zulfiya Davlyatnazarova (Speaker), Institute of Botany, Plant Physiology, and Genetics
- Azizakhon Haidarova, International Science and Technology Center, Tajikistan Branch Office
- Khurram Khayrov, E.N. Pavlovsky Institute of Zoology and Parasitology
- Mukhabatsho Khikmatov, International Science and Technology Center, Tajikistan Branch Office
- Mehri Rustamova, Tajikistan National Academy of Sciences

Taiwan

- Ya-Hsin Li (Speaker), Chung-Shan Medical University

Uganda

- Hellen Nansumba (Speaker), Ministry of Health of the Republic of Uganda

United States

- Faina Linkov (Committee Member), Duquesne University
- Taunton Paine (Speaker), National Institutes of Health
- Lusine Poghosyan (Committee Chair), Columbia University
- Vasiliki Rahimzadeh (Committee Member), Baylor College of Medicine
- Krystal Tsosie (Committee Member), Arizona State University
- Gautham Venugopalan (Committee Member), Gryphon Scientific
- Wei Zheng (Committee Member), Vanderbilt University Medical Center

Uzbekistan

- Ravshan Azimov (Speaker), Tashkent Medical Academy
- Shakhlo Turdikulova (Speaker), Uzbekistan Ministry of Higher Education, Science, and Innovation

Appendix E
Commissioned Papers

The authors are solely responsible for the content of these papers, which do not necessarily represent the views of the National Academies of Sciences, Engineering, and Medicine.

Global Biodiversity Information Facility (GBIF): Kick-Starting the Biodiversity Data Publication Process for Tajikistan

Samariddin Barotov, Senior Researcher at the Institute of Botany, Plant Physiology, and Genetics, National Academy of Sciences of Tajikistan, and Node Manager for GBIF in Tajikistan

Creating a Biological Data Framework: The Global Biodiversity Information Facility

In 1992, the United Nations Conference on Environment and Development, also known as the Rio Earth Summit, culminated in the signing of the Convention on Biological Diversity (CBD). This institutional framework was created for the protection of biological diversity (Bell, 1993). By promoting the sustainable use and equal benefits of sharing genetic materials, its provisions were designed to facilitate better access to genetic resources (Panjabi, 1993). The use of biological resources requires data repositories and distribution nodes, which are a type of resource center that enables the preservation and distribution of biological materials and data information. These biological centers are essential for research and development in the life sciences, and they serve many critical roles, including the preservation of biological resources and biodiversity conservation.

In Tajikistan, a prime example of a biological resource center is the Global Biodiversity Information Facility (GBIF), an international network and data initiative that provides free and open access to biodiversity data (GBIF, 2021). The GBIF network consists of more than 100 participating countries and organizations, including countries in Central Asia, the United States, and participating member states of the European Union. Working through the participant nodes, GBIF is an international coordinating body created to promote and enable the global dissemination and use of the world's biodiversity data, and to provide data holding institutions with common standards and open-source tools. Using data standards such as those established by the Taxonomic Databases Working Group (n.d.) and the International Union of Biological Sciences (n.d.), GBIF has access to hundreds of millions of biological species occurrence records.

These standards provide straightforward and collaborative frameworks and platforms for scientists from many disciplines and nationalities for promoting research, training, and educational opportunities in the life sciences. In this way, the GBIF network provides scientific researchers with open access to biological data. The data are made accessible under the Creative Commons License, permitting researchers to adapt the data in peer-reviewed publications and policy papers, provided the original work and source are cited appropriately. Many of the life sciences publications from the Central Asia region—which cover a wide range of topics from the impacts of climate change to the spread of invasive and alien pests to priorities for conservation, food security, and human health—would not be possible without the data provided by the GBIF network (GBIF, 2021).

The Use of GBIF in Central Asia: Tajikistan

Located in the Central Asia region, Tajikistan has a diverse array of flora and fauna, contributing to its rich biodiversity. Given the importance of agricultural crops in the region, a number of measures, programs, and initiatives have been developed to facilitate open access to biological data, operating in accordance with governing laws, conservation, and management (Kotowski, 2022). One such example is the national law on "Conservation and Sustainable Use of Crop Genetic Resources," which was developed in 2012 to better preserve wild plant species in protected areas (Turok, 2013). This law was also intended to further support and protect farmers and their activities and rights around local plant diversity efforts, as well as access to and benefits sharing of plant genetic resources (Turok, 2013). In its totality, this law helped to maintain the agricultural industry in Tajikistan, ensuring food, environmental, and biological security; enabling scientific research and development; and safeguarding sociocultural and historical heritage for the prosperity of both current and future generations.

Regarding bioprospecting and access to biological data, Tajikistan currently faces limited regulations concerning genetic data resources. Uncertainties exist about the role of biological resources and their rational utilization. Nonetheless, several projects in the country receive support from international organizations, including the Food and Agriculture Organization of the United Nations (FAO), the United Nations Environment Programme, the CBD, and the Global Environment Facility (GEF) (FAO, 2022; Nasyrova, 2011). For example, in 2021, the GEF and FAO colaunched a project in Tajikistan to improve regulations surrounding the use of agrobiodiversity in a 3-year project titled Facilitating Agrobiodiversity Conservation and Sustainable Use to Promote Food and Nutritional Resilience in Tajikistan. This project focused on strengthening the country's nationally and globally significant biodiversity by maintaining local plants and wildlife crops (FAO, 2022). In addition, the project aimed to create gene banks in specific areas of the region, while ensuring equitable distribution of benefits (FAO, 2022).

The governance of biological data in Tajikistan is primarily overseen by various governmental organizations, such as the Ministry of Agriculture, the Committee for Environmental Protection, the Ministry of Health of Tajikistan, the National Academy of Sciences of Tajikistan, and the Academy of Agricultural Sciences of Tajikistan (WHO, 2020). However, the digitalization of data remains limited, impacting the accessibility of medical and agricultural data in the country. Additionally, the sharing of genetic data presents many challenges as well. According to national law in Tajikistan, for these data types to be shared, one must sign a memorandum of understanding or agreement with the recipient organization. This must follow the Nagoya Protocol on Access and Benefit-Sharing, which Tajikistan signed on September 21, 2011 (Kamau et al., 2010). Under the Nagoya Protocol, participating countries aim to equitably share the benefits arising from the use of genetic resources (Kamau et al., 2010).

While Tajikistan has made strides in regulating and increasing the accessibility of biological data, there are ongoing efforts to address current inadequacies, as not all biological data are accessible for open sharing. Additionally, because it is a young country, there are pressures to develop faster. To keep pace with the rapidly advancing fields of science, technology, and medicine, actions to digitalize data, establish local platforms for data management, use artificial intelligence for data regulation and management (e.g., OpenAI's ChatGPT), train young specialists in bioinformatics and bioengineering, provide necessary equipment and laboratory facilities, and promote data transparency are important. Additionally, in the field of medicine, Tajikistan obtains a lot of data; however, it is unfortunate that most of these datasets are not digitized for public users, especially in the Tajik language. Ultimately, Tajikistan, along with other Central Asian countries, should prioritize capacity building in data regulation and management, implement proper data governance methodologies, secure funding support, and foster collaboration among institutions. Cooperative efforts will result in the consolidation of substantial, high-quality data for open sharing, fostering scientific advancements and sustainable development.

Disclaimer: The author is solely responsible for the content of this paper, which does not necessarily represent the views of the U.S. National Academies of Sciences, Engineering, and Medicine.

Edited by: Carmen Shaw, U.S. National Academy of Sciences, Engineering, and Medicine.

References

Bell, D. E. 1993. The 1992 Convention on Biological Diversity: The continuing significance of U.S. objections at the Earth Summit. *George Washington Journal of International Law and Economics* 26(3):479-538.

FAO (Food and Agriculture Organization of the United Nations). 2022. FAO, GEF promote agrobiodiversity and sustainability to improve resilience in Tajikistan. Blog. *Family Farming Knowledge Platform.* https://www.fao.org/family-farming/detail/en/c/1606372 (accessed January 24, 2024).

GBIF (Global Biodiversity Information Facility). 2021. *Strategic framework 2023-2027.* GBIF Secretariat: Copenhagen. https://doi.org/10.35035/doc-0kkq-0t82.

International Union of Biological Sciences. n.d. *International Union of Biological Sciences (IUBS).* https://iubs.org (accessed August 4, 2023).

Kamau, E., B. Fedder, and G. Winter. 2010. The Nagoya Protocol on Access to Genetic Resources and Benefit Sharing: What is New and what are the implications for provider and user countries and the scientific community. *Law, Environment and Development Journal* 6(3):246.

Kotowski, M. A., S. Świerszcz, C. K. Khoury, M. Laldjebaev, B. Palavonshanbieva, and A. Nowak. 2022. The primal garden: Tajikistan as a biodiversity hotspot of food crop wild relatives. *Agronomy for Sustainable Development* 42:Article 112. https://doi.org/10.1007/s13593-022-00846-9.

Nasyrova, F. 2011. Legal aspects of bioethics in Tajikistan. In *Genomics and bioethics: Interdisciplinary perspectives, technologies and advancements.* Hershey, PA: IGI Global. Pp. 220-234.

Panjabi, R. K. L. 1993. International law and the preservation of species: An analysis of the Convention on Biological Diversity signed at the Rio Earth Summit in 1992. *Dickinson Journal of International Law* 11(2):187-282.

Taxonomic Databases Working Group. n.d. *Biodiversity information standards (TDWG).* https://www.tdwg.org (accessed August 4, 2023).
Turok, J., M. Begmuratov, K. Akramov, C. Carli, S. Christmann, M. Glazirina, K. Jumaboev, A. Karimov, J. Kazbekov, Z. Khalikulov, R. Mavlyanova, N. Nishanov, A. Nurbekov, N. Saidov, R. Sharma, K. Toderich, M. Turdieva, and T. Yuldashev. 2013. *Agricultural research collaboration in Tajikistan.* Working Paper No. 14. Beirut, Lebanon: International Center for Agricultural Research in the Dry Areas. https://hdl.handle.net/20.500.11766/7901 (accessed October 15, 2021).
WHO (World Health Organization). 2020. *Joint external evaluation of IHR core capacities of the Republic of Tajikistan: Mission report, 21-25 October 2019.* License: CC BY-NC-SA 3.0 IGO. Geneva: World Health Organization.

Data Governance Experience in Epidemiological Studies in Kazakhstan: Insights and Implications

Alexandr Ivankov
Independent Researcher, Almaty, Kazakhstan

Introduction

The rapid rise in the volume of data is driving the importance of analyzing data governance policies related to the health care sector (Raghupathi et al., 2014). Global practice of data governance is characterized by the widespread introduction of data management standards at the interstate and national levels. Additionally, issues related to personal data security in health care are of great sensitivity and significance. In recent years, Kazakhstan has been actively developing the principles of open data governance by expanding the range of data available for open use, including the possibility of open data application programming interface (API), where different end users can use open data provided by private organizations and/or the government.[1] This technology allows developers to utilize ready-made blocks to build their own applications. Currently, open API is successfully employed on the eGov Open Data portal, which provides a single mechanism for interactions between the state, citizens, and other government agencies (eGov, n.d.). The main purposes of this paper are to describe the process of data governance in epidemiological research, identify related problems, and suggest possible development options for Kazakhstan.

Results

The research methodology employed in this paper involved an analysis of the data management and data governance practices utilized in a series of COVID-19 epidemiological studies conducted in Kazakhstan (Dyusupova et al., 2021; Semenova et al., 2020a, 2022). The analysis included a description of personal experiences, internal documentation, and national data governance legislation that ensured the confidentiality of personal data in Kazakhstan. Some of these legislations include the laws titled On Approval

[1] Approval of the Concept of Digital Transformation, Development of the Industry of Information and Communication Technologies and Cyber Security for 2023–2029. *Decree of the Government of the Republic of Kazakhstan.* No. 269 (2023) (see https://adilet.zan.kz/rus/docs/P2300000269).

of the Rules for Conducting Clinical Trials of Medicines and Medical Devices,[2] On Approval of the Rules for Conducting Biomedical Research,[3] and On Approval of Good Pharmaceutical Practices.[4] The methodology employed focused on the understanding of data governance in epidemiological research within Kazakhstan, specifically in the context of the COVID-19 pandemic. The approach included an examination of national legislation and recent laws introduced in 2023 regarding data governance in Kazakhstan.

Additionally, this paper analyzes data management and governance practices employed in various COVID-19 epidemiological studies conducted in the country. This includes studies on disease severity and mortality in COVID-19 patients with diabetes mellitus. Existing procedures that regulate data storage and processing (Dyusupova et al., 2021). The process of data storage and processing was characterized by the classification of data into categories *significant* and *not significant*, *requiring confidentiality* or *public access*, followed by data storage and a description of processing methods, including statistical analysis in accordance with the study protocol approved by the ethical committee. For example, in the case of a COVID-19 outbreak forecast study (Semenova et al., 2020b), the data were classified as *public* and were freely discussed and exchanged within the team. When working on the COVID-19-diabetes impact study (Dyusupova et al., 2021), the data were available only after being processed by the anonymization group, which had previously been trained on relevant methods of working with the data, and subsequently stored on university computers. The process of monitoring compliance with the protocol for working with data was carried out by the local committee in accordance with the internal committee processes.

The methodology emphasized the creation of internal, unique procedures for each study. Collaboration among researchers was a central part of the study, involving institutional partnerships with the Medical University of Semey City and the In Vitro Laboratory based at Al-Farabi Kazakh National University for large-scale studies using big data. Ethical considerations were also at the forefront, with specific regulations and agreements regarding data handling adapted to the unique conditions of the pandemic. Changes in data management policies are proceeding quickly, and some of the problems have already been eliminated at the legislative level with the introduction of a new 2023 law on data governance, titled the Law on Personal Data and Their Protection (Iskakova and Kadyrzhanova, 2022). Furthermore, the general principles outlined in this law, along with the implementation of the "Data Drive State" strategy and the current e-government efforts on open data, inspire optimism that these trends will also spread to data from the health care sector, thereby increasing the scientific potential for future research, optimizing

---

[2] On Approval of the Rules for Conducting Clinical Trials of Medicines and Medical Devices, Clinical and Laboratory Tests of Medical Devices for Diagnostics Outside a Living Organism (in vitro) and the Requirements for Clinical Bases and the Provision of Public Services, Medicinal Products, and Medical Devices. Order of the Minister of Health of the Republic of Kazakhstan No. 21772 (2020) (see https://adilet.zan.kz/rus/docs/V2000021772).

[3] On Approval of the Rules for Conducting Biomedical Research and Requirements for Research Centers. Order of the Minister of Health of the Republic of Kazakhstan No. 21851 (2020) (see https://adilet.zan.kz/rus/docs/V2000021851).

[4] On Approval of Good Pharmaceutical Practices. Acting Order Minister of Health of the Republic of Kazakhstan No. 22167 (2021) (see https://adilet.zan.kz/rus/docs/V2100022167).

the management of the health care sector, and improving the security of personal data.[5] All ethical considerations, along with the prerequisite of registering studies involving human subjects, are broadly outlined in national legislation. However, these laws contain only general points on data governance in research, therefore necessitating the creation of a new operational procedure to establish a logical pathway for data collection, safety, storage, processing, and analysis in epidemiological studies.

During the COVID-19 pandemic, this author's research group conducted several studies, including two studies based on analysis of publicly available data from Our World in Data, an online vaccination dataset, and reports from the National Center for Public Health of the Ministry of Health of Kazakhstan (Mathieu et al., 2020). The unique conditions arising during the pandemic introduced adaptations to the traditional processes of data collection and processing methodologies that were in place before. In the initial two epidemiological studies on the characteristics of coronavirus infection in Kazakhstan and predictive modeling of its spread, publicly available datasets, updated daily, were utilized, thereby fostering proximity to real-time insights for subsequent analytical and research pursuits. The greatest challenge was the inability to copy data from open and local sources in Kazakhstan due to their presentation in noncopyable PDF formats. This resulted in the need for manual data transfer over the course of several months, consequently impeding the speed of work and necessitating the addition of an extra assistant to the data entry team. Typically, data entry tasks are handled by two trained operators; however, due to limited data accessibility and a high risk of systematic input errors, additional assistants were brought in for these studies to oversee data entry and verification.

It is noteworthy that the execution of these studies occurred within a specific ethical framework, permitting an exemption from the usual requirement of study registration in cases not involving interventions. This was primarily driven by the need to expedite obtaining data on the behavior of the coronavirus infection and its management pathways. In the study dedicated to examining the clinical characteristics and risk factors for disease severity and mortality in COVID-19 patients with diabetes mellitus in Kazakhstan, uniquely, data collected by the Ministry of Health of Kazakhstan during the initial waves of the pandemic were analyzed (Dyusupova et al., 2021). The collected information, in the form of excerpts from medical records with the complete removal of any identifying information, such as the individual identification number of citizens of Kazakhstan and their full names, was obtained in raw form by the Medical University of Semey City, recognized as the most experienced in conducting large-scale epidemiological studies in the country (Dyusupova et al., 2021). This is because of the university's expertise in research related to the impact of adverse environmental conditions and the consequences of nuclear testing at the Semipalatinsk nuclear test site. However, because of unprecedented conditions during the pandemic, the absence of established standard procedures for data transmission and collaborative work, and the unique nature of the study, an internal directive was issued that regulated the university staff's rights to access the data, the rights to authorship and coauthorship, and restrictions on data sharing with third parties. All university staff members signed a form expressing their agreement to maintain information security, refrain from sharing with third parties, and use the data solely for scientific purposes using only password-protected office computers.

[5] About Access to Information. Law of the Republic of Kazakhstan No. 401-V ZRK (2015) (see https://adilet.zan.kz/rus/docs/Z1500000401/z150401.htm).

Another large-scale study using big data was conducted by this author's research team in collaboration with the *IN VITRO* Laboratory based at Al-Farabi Kazakh National University (Semenova et al., 2022). Since there was no preexisting standard operating procedure for obtaining and processing large, anonymized datasets from laboratories, an internal procedure was developed that enabled work in a secure mode of data storage and analysis. Within this procedure, data transmission and storage were restricted to an office computer with a password that changed twice a week and was known only to the principal investigator and the data analyst. Unlike the study Clinical Characteristics and Risk Factors for Disease Severity and Mortality of COVID-19 Patients with Diabetes Mellitus, for which data were obtained as a database rather than primary information, the concern about accidental inclusion of identifying data did not arise. However, there was an issue regarding which computer to use for data analysis, as office computers lacked the processing power required for the dataset of 85,346 observations. Consequently, an additional computer was specifically acquired and documented as one of the key components in the technical infrastructure of the study.

During the conduct of all studies, existing procedures that regulate data storage and processing were also examined, providing insights into the general principles of working with data. The creation of internal procedures unique to each study was urgently required. The lack of a previous standardized approach was mainly due to the uniqueness of the pandemic conditions. Simultaneously, challenges such as low data availability, the need to rely on internationally reported data, manual transfer of aggregated publicly available data digitized by the state, and the absence of technical infrastructure for handling big data posed significant obstacles to the speed of conducting research and publishing results. Based on the results of the analysis, it was found that at all levels—national legislation, internal policies of organizations, and data exchange protocols prescribed for a concrete study and approved by the local ethical committee (as part of the approval of the general study protocol)—there are elements that correspond to the norms of international practices. However, the role of supervisory authorities includes prosecutorial authorities, which exercise the highest supervision over the observance of legality in the field of personal data and their protection, as well as the central and local ethics committees, which serve as internal bodies for data control. The functionality of these entities varies among different committees, and it is unclear in the epidemiological studies who provides control over the storage of data and compliance with the principles specified in the relevant documents, along with how they exercise this control.

Some problems identified are the difficulty of accessing open government and medical health data and the quality of the published data. The lack of understanding of how and under what conditions it is possible to access data from health care organizations limits the prospects of obtaining new scientific knowledge. The lack of understanding of who controls the security of data obtained during the research and how they control it inspires concern for the security of these data.

Based on the results of the studies analyzed and current data control practices in health care research, some recommendations might be suggested:

1. Develop Clear Guidelines and Protocols:
    a. Establish transparent and standardized guidelines for accessing government health data and data from large national medical projects.

   b. Create protocols that clearly define how and under what conditions researchers and organizations can access health care data.
2. Enhance Data Quality:
   a. Implement strict quality control measures to ensure published data are accurate, consistent, and reliable.
   b. Develop a centralized system for auditing and verifying the quality of data being shared.
3. Educate and Train Stakeholders:
   a. Provide education and training to researchers, policymakers, and other stakeholders to enhance their understanding of data access, quality, and security protocols.
   b. Organize workshops and training sessions to educate about responsible data access and use procedures.
4. Facilitate Collaboration and Communication:
   a. Foster collaboration among government agencies, research institutions, and health care organizations to facilitate seamless access to health data.
   b. Develop communication channels to ensure that concerns and issues related to data access and security are promptly addressed.
5. Build Public–Private Partnerships:
   a. Engage with private-sector organizations that have expertise in data management and security.
   b. Leverage these partnerships to develop innovative solutions for data access, quality, and security challenges.
6. Promote Transparency:
   a. Maintain transparency in the process of data collection, storage, and sharing.
   b. Publicize the procedures and protocols to enhance public trust and encourage responsible usage.
7. Utilize Technology and Innovation:
   a. Develop secure platforms and interfaces that allow seamless access for researchers, while maintaining compliance with ethical and legal standards.
8. Evaluate and Adapt Regularly:
   a. Regularly assess the effectiveness of the implemented solutions.
   b. Make necessary adjustments and refinements to keep the strategies aligned with evolving needs and challenges.

By addressing these areas, it is possible to build a more open, reliable, and secure ecosystem for accessing and utilizing health care data, thereby enhancing the prospects for obtaining new scientific knowledge and maintaining the security of the data.

Conclusion

This paper contributes to the ongoing discourse on data governance in epidemiological research and provides a foundation for future advances in data management practices. It underscores the significance of aligning local data management policies with international standards, fostering responsible data governance, and promoting efficient public health policies in Kazakhstan.

Disclaimer: The author is solely responsible for the content of this paper, which does not necessarily represent the views of the U.S. National Academies of Sciences, Engineering, and Medicine.

References

Dyusupova, A., R. Faizova, O. Yurkovskaya, T. Belyaeva, T. Terekhova, A. Khismetova, A. Sarria-Santamera, D. Bokov, A. Ivankov, and N. Glushkova. 2021. Clinical characteristics and risk factors for disease severity and mortality of COVID-19 patients with diabetes mellitus in Kazakhstan: A nationwide study. *Heliyon* 7(3).

eGov. n.d. *Government services and information online of the Republic of Kazakhstan.* https://egov.kz/cms/ru (accessed August 14, 2023).

Iskakova, Z. T., and T. S. Kadyrzhanova. 2022. Analysis of problems and challenges in the legislation of the Republic of Kazakhstan on personal data protection and international legal regulation. *Bulletin of L. N. Gumilyov Eurasian National University Law Series* 4(141):49-60.

Mathieu, E., H. Ritchie, L. Rodés-Guirao, C. Appel, C. Giattino, J. Hasell, B. Macdonald, S. Dattani, D. Beltekian, E. Ortiz-Ospina, and M. Roser. 2020. Coronavirus pandemic (COVID-19). *Our World in Data.* https://ourworldindata.org/coronavirus (accessed January 24, 2024).

Raghupathi, W., and V. Raghupathi. 2014. Big data analytics in healthcare: Promise and potential. *Health Information Science and Systems* 2:1-10.

Semenova, Y., N. Glushkova, L. Pivina, Z. Khismetova, Y. Zhunussov, M. Sandybaev, and A. Ivankov. 2020a. Epidemiological characteristics and forecast of COVID-19 outbreak in the Republic of Kazakhstan. *Journal of Korean Medical Sciences* 35(24):e227. https://doi.org/10.3346/jkms.2020.35.e227.

Semenova, Y., L. Pivina, Z. Khismetova, A. Auyezova, A. Nurbakyt, A. Kauysheva, D. Ospanova, G. Kuziyeva, A. Kushkarova, A. Ivankov, and N. Glushkova. 2020b. Anticipating the need for healthcare resources following the escalation of the COVID-19 outbreak in the Republic of Kazakhstan. *Journal of Preventive Medicine and Public Health* 53(6):387-396. https://doi.org/10.3961/jpmph.20.395.

Semenova, Y., Z. Kalmatayeva, A. Oshibayeva, S. Mamyrbekova, A. Kudirbekova, A. Nurbakyt, A. Baizhaxynova, P. Colet, N. Glushkova, A. Ivankov, and A. Sarria-Santamera. 2022. Seropositivity of SARS-CoV-2 in the population of Kazakhstan: A nationwide laboratory-based surveillance. *International Journal of Environmental Research and Public Health* 19(4):2263. https://doi.org/10.3390/ijerph19042263.

Process of Developing a Center for Applied Artificial Intelligence and Cyber Security at the Kyrgyz State Technical University in Kyrgyzstan

Gulnara J. Kabaeva, Director, Institute of Information Technologies of I. Razzakov Kyrgyz State Technical University, Doctor of Physical and Mathematical Sciences, Professor in Computer Science (certified by Higher Attestation Commission of Kyrygyzstan), Bishkek, Kyrgyzstan

This article discusses the issues of data management for information technology (IT) companies in Kyrgyzstan, along with tasks accomplished by the Center for Applied

Artificial Intelligence and Cybersecurity at the I. Razzakov Kyrgyz State Technical University. (KSTU).

Introduction

Modern times have become driven by the digital economy and propelled by transformative IT innovations. This has marked the fourth industrial revolution, with intellectual technologies being developed rapidly and used in almost all areas of human activity, including medicine and agriculture. The development and use of intelligent technologies were built on the processing of large datasets. This has led to new attitudes toward data and to the development of policies and strategies for data management, both in individual organizations and at the level of governmental decisions.

For the digital transformation of the Kyrgyzstan economy, an action plan was developed to implement the digitalization of state and municipal government operations, titled the National Development Programme of the Kyrgyzstan until 2026. This plan included the development and management of digital infrastructures, the provision of high-quality digital services, the development of IT education, and the training of highly qualified IT specialists for the industry.[6] The implementation of the plan includes improving the regulatory framework for data management, including the collection, storage, use, and security of data. The data management strategy is aimed at optimizing the management of companies and their activities, employees, and connected devices, as well as within countries—with policies that are compliant with national and international data frameworks (Dama International, 2017).

Kyrgyzstan's burgeoning IT sector has been evident in recent years. However, not all enterprises are confined to software development, with other services encompassing IT services, hardware vendors, and equipment component parts. This growth is noted by news outlets, such as tazabek.kg, which provides a list of 80 IT companies (*Tazabek*, n.d.). There is an absence of large domestic IT companies in Kyrgyzstan currently, which is mitigated by export-related company activities. In 2011, the High Technology Park of Kyrgyzstan was founded, which unites more than 200 IT companies. The park provides its residents with special tax and legal regimes that exempt export-oriented IT companies from certain types of taxes.[7]

Digital data management in software companies includes a wide range of tasks, behind-the-scenes application of policies and actions that ensure competent and secure work with data, and a choice of platforms for implementing their solutions. To analyze the data management strategies of IT companies, an internal audit should be carried out, if necessary. Well-known data management work steps include generating data, data storage, making data publicly available, updating data, and recovering from system failures or

---

[6] National Development Program of the Kyrgyz Republic until 2026. Decree of the President of the Kyrgyz Republic. No. 435 (2021) (see https://cbd.minjust.gov.kg/430700?refId=1096469); Action Plan for Digitalization of Management and Development of Digital Infrastructure in the Kyrgyz Republic for 2022-2023. Order of the Cabinet of the Ministers of the Kyrgyz Republic. No. 2-r (2022) (see http://cbd.minjust.gov.kg/act/view/ru-ru/218797).

[7] Regulations on the Procedure for Registering Residents of the High Technology Park of the Kyrgyz Republic. Decree of the Government of the Kyrgyz Republic. No. 267 (2012) (see http://cbd.minjust.gov.kg/act/properties/ru-ru/93604/10); On the High Technology Park of the Kyrgyz Republic. Law of the Kyrgyz Republic. No. 84 (2011) (see http://cbd.minjust.gov.kg/act/view/ru-ru/203327).

emergencies, as well as utilizing data in certain applications and computing tasks, and ensuring the confidentiality and security of data.

Regulatory Framework for Data Management in the Kyrgyzstan

Data companies operating in Kyrgyzstan must familiarize themselves with and adhere to the existing regulatory and legal provisions and laws of the country. The list of such documents includes[8]:

- The Constitution of Kyrgyzstan.
- Law of Kyrgyzstan, On Personal Information.
- Law of Kyrgyzstan, On e-Governance.
- Law of Kyrgyzstan, On Personal Data Protection.
- As amended by the Laws of Kyrgyzstan, About Trade Secrets.
- Decree of the Government of Kyrgyzstan, Cybersecurity Strategy of Kyrgyzstan for 2019-2023.
- Decree of the Government of Kyrgyzstan, On Approval of the Requirements for the Protection of Information Contained in Databases of State Information Systems.
- Order of the Cabinet of Ministers of Kyrgyzstan, On Approval of the Concept of Open Data of Kyrgyzstan for the Period 2022-2024.

One can become acquainted with the content of these documents on the Open Data Portal of Kyrgyzstan, which has accessible legal information for various sectors, including education, health care, transport, etc. Additionally, in September of 2022, the government approved the Open Data Concept of Kyrgyzstan, making the country one of the first in Central Asia to join the Open Government Partnership. Kyrgyzstan ranks 58th in the world for level of accessibility of digital content.[9]

It is noteworthy to mention that the Personal Data Protection Law of Kyrgyzstan is in line with the European Union's General Data Protection Regulation.[10] Within it, the seven key principles for the management and processing of personal data include legality, fairness, transparency, accuracy, integrity, and confidentiality, as well as compliance with restrictions and storage requirements (ISO, 2022a,b). On the website of the State Agency for the Protection of Personal Data (2023) under the Cabinet of Ministers of Kyrgyzstan, there are regulatory documents that relate to the issues of ensuring the protection of human rights and freedoms related to collecting, processing, and using personal data.

Difficulties associated with data management primarily arise from the ever-increasing volumes of data that have different forms of representations and types. Data can

---

[8] See Regulatory and Legal Provisions and Laws of the Kyrgyz Republic following the references list for this paper.

[9] Open Data Concept for Kyrgyz Republic for the period 2022-2024. Order of the Cabinet of Ministers of the Kyrgyz Republic No. 463-r (2022) (see http://cbd.minjust.gov.kg/act/view/ru-ru/219184?cl=ru-ru).

[10] On the protection of natural persons with regard to the processing of personal data and on the free movement of such data, and repealing Directive 95/46/EC (General Data Protection Regulation). Regulation (EU) 2016/679 of the European Parliament and of the Council (2016) (see https://eur-lex.europa.eu/legal-content/EN/TXT/?uri=CELEX%3A02016R0679-20160504&qid=1532348683434).

come from information devices such as sensors, video cameras, medical devices, internet sources, and electronic measuring instruments. In addition, the received data require storage and modern, high-speed data management systems, and backup storage units. As a result, companies must adhere to the requirements of data confidentiality related to national security and international laws, so as not to violate copyrights and property rights. In this way, copyright protection is provided by the State Agency for Intellectual Property and Innovations of Kyrgyzstan[11] (Kyrgyzpatent, 2023), which ensures application examinations and issues protection titles in accordance with the legislation of Kyrgyzstan in the field of intellectual property.

Possible Risks and Vulnerabilities in the Kyrgyzstani IT Sector

For actively operating IT companies, there are and always will be risks related to data, as well as processes that may affect the implementation and use of data-related tasks. Undoubtedly, risks associated with software development in the IT sector can lead to both reversible and irreversible consequences, such as information leakage and data loss, personnel errors, and confidentiality violations related to standards and norms for risk management. To prevent and reduce human-induced errors, careful training and work with personnel are necessary. Additionally, IT companies in Kyrgyzstan are engaged in various tasks related to the automation of technological processes, the management of electronic documentation, intelligent solutions in the financial sector, web applications, and automated information systems in educational and medical institutions, as well as e-commerce support systems (ISO, 2009). There are companies that are distributors of hardware and software from large international companies. In addition, cellular operators have their own software development departments, and there are state-owned IT enterprises under the Ministry of Digital Development of Kyrgyzstan that are working on digital solutions for the country's economy.

Software development is always associated with data processing, data arrays, and competent work with data. With each passing year, the amount of data will accumulate, and the problem of secure data storage will require solutions. For companies to stay in a competitive market, they must adhere to the rules of data protection and comply with international intellectual property and ethical legal standards. Companies that work for export gain access to customer databases and are required to comply with the rules and requirements of the security policy and international data and risk management standards. They define their relationship through contracts, cooperative agreements, and nondisclosure agreements for trade secrets. The presence of international standards greatly contributes to good data management, data quality management planning, and organization risk mitigation.

Tasks of the I. Razzakov Institute of Information Technologies of KSTU

Innovative artificial intelligence technologies based on the development of both software and computer technology have increased in demand for the IT market. IT companies in Kyrgyzstan are potential employers for graduates of the republic's universities. As is the case all over the world, the demand for IT specialists in Kyrgyzstan

---

[11] On the Issues of the State Agency Intellectual Property and Innovation. Resolution of the Cabinet of Ministers of the Kyrgyz Republic. No. 111 (2021) (see http://patent.gov.kg/).

is high, and, like everywhere else, developers with knowledge and skills in the latest technologies are needed. The demands of the domestic labor market can be met mainly by graduates of local universities. The I. Razzakov Kyrgyz State Technical University (KSTU) is the only engineering institution of higher education in Kyrgyzstan; KSTU has been training IT specialists in a wide range of specialties since the 1980s. However, the way to remain in the educational services market, where IT companies are oftentimes looking for qualified specialists, is the modernization of educational programs in accordance with the requirements of the IT companies and other emerging global trends.

The Institute of Information Technologies at KSTU offers 11 undergraduate and 6 master's degree programs in IT areas (IIT, 2023). In 2021, a new undergraduate educational program in big data analytics was launched to provide the domestic market of the country with the missing specialists in the field of data analysis and processing, given the high demand in IT. This program began because of the implementation of the Erasmus + Coordinating Board for Higher Education project: "Establishment of Training and Research Centers and Courses Development on Intelligent Big Data Analysis in CA/ELBA, 2019-2023" (KSTU, 2023). This project created the Laboratory of Data Analysis, equipped with modern computers, with professors trained in European universities of the project partners. KSTU also held training in the field of big data mining with access to highly efficient modern tools based on the experience and technologies of the European Union.

Kyrgyzstan is 94th on the Global Innovation Index 2022. Indicators of low levels of innovation links (13.7%), research collaboration between universities and industry (24.3%), and knowledge impact (15.1%), as well as the need to train specialists with knowledge of artificial intelligence methods and technologies, led to the decision to create the Center for Applied Artificial Intelligence and Cyber Security as the most relevant and knowledge-intensive field today. The goals are to improve the quality of educational services; improve the educational programs of KSTU in IT areas, while regulating educational results; and maintain feedback with employers. Additionally, the development of IT education with the implementation of the education–science–business model, while training highly qualified IT personnel and thereby increasing the employment potential of graduates of IT specialties of KSTU in modern conditions, are also highly prioritized. Thus, the Center for Applied Artificial Intelligence and Cyber Security is designed to study and implement:

- Practical application of artificial intelligence in various fields, including education, health care, emergency situations, and energy.
- Development of new business software products.
- New educational programs for the preparation of bachelor's and master's degrees in artificial intelligence and machine learning in Western universities under dual degree programs.

As part of the center, laboratories are being created to develop:

- Artificial intelligence methods and technologies for applied research problem-solving in the fields of science and technology.
- Industrial automation.
- Cybersecurity and information security analysts.

- The Internet of Things.

IT teachers and students are developing the university's learning management system, as well as the Advisement Verification Number automated information system, originally created at the university. Institute teachers and students carry out scientific work using methods and algorithms of artificial intelligence to study:

- Software and hardware systems in the field of computer vision and augmented and virtual reality.
- Natural language processing technologies on IT (work is underway in this field with the company ULUT soft).
- Recommendatory and intelligent decision support systems in education and medicine.
- Development of digital solutions for health care and social organizations.
- Information security systems for the banking sector.
- Development of a monitoring system with the help of intelligent recognition of the danger of a breakthrough in high mountain lakes.

Conclusion

The digitalization of the economy and the use of intelligent technologies based on the processing of large datasets have led to the need to develop both policies and strategies for data management in individual organizations and the government. In Kyrgyzstan, for the implementation of digital solutions, a legal and regulatory framework related to data management (collection, cleaning, storage, use, and security) is being developed consistently. The number of IT companies is growing in the country, most of which work for export under foreign contracts. When working with data, it is necessary to resolve issues related to data quality, data protection, and copyright, based on international standards. Needs include training specialists with knowledge of the methods and technologies of artificial intelligence at KSTU, while considering the need to regulate educational results, and maintain feedback with the employer-led decision to create a Center for Applied Artificial Intelligence and Cyber Security.

Disclaimer: The author is solely responsible for the content of this paper, which does not necessarily represent the views of the U.S. National Academies of Sciences, Engineering, and Medicine.

References

Dama International. 2017. *DAMA-DMBOK: Data management body of knowledge.* Technics Publications, LLC.

IIT (Institute of Information Technologies) of KSTU named after I. Razzakova. n.d. https://kstu.kg/en/bokovoe-menju/faculties/institute-of-information-technology (accessed February 13, 2024).

ISO (International Organization for Standardization). 2009. *Risk management—Principles and guidelines.* https://www.iso.org/obp/ui/#iso:std:iso:31000:ed-1:v1:en (accessed January 24, 2024).

ISO. 2022a. *Data quality. Part 1: Overview*. ISO 8000-1. https://www.iso.org/obp/ui/#iso:std:iso:8000:-1:ed-1:v1:en (accessed January 24, 2024).
ISO. 2022b. *Data quality. Part 2: Vocabulary*. ISO 8000-2. https://standards.iteh.ai/catalog/standards/iso/2e74ce51-74cf-4b5e-8089-295dfafcb8a2/iso-8000-2-2022 (accessed January 24, 2024).
Kyrgyzpatent. n.d. Official Site of Kyrgyzpatent. http://patent.gov.kg (accessed August 25, 2023).
The State Personal Data Protection Agency. *What is personal data?* https://dpa.gov.kg/en (accessed August 25, 2023).
KSTU (Kyrgyz State Technical University named after I. Razzakova). n.d. *ELBA Project*. https://kstu.kg/proekty/tekushchie/proekt-modernizacija-vysshego-obrazovanija-v-centralnoi-azii-cherez-novye-tekhnologii-hiedtec-po-programme-ehrazmus-1 (accessed August 20, 2023).
Tazabek. n.d. 80 It-Компаний Кыргызстана - Владельцы и Учредители - Tazabek. https://www.tazabek.kg/news:1929262 (accessed August 7, 2023).

Regulatory and Legal Provisions and Laws of the Kyrgyz Republic

Constitution of the Kyrgyz Republic. Law of the Kyrgyz Republic. (2021) (see http://cbd.minjust.gov.kg/act/view/ru-ru/112213/10?mode=tekst).
On Personal Information. Law of the Kyrgyz Republic. No. 58 (2008) (see https://dpa.gov.kg/en/npa/4).
On Amendments to the Law of the Kyrgyz Republic "On Personal Information." No. 129 (2017) (see http://cbd.minjust.gov.kg/act/view/ru-ru/111636?cl=ru-ru).
On Approval of the Procedure for obtaining the consent of the subject of personal data to the collection and processing of his personal data, the procedure and form for notifying subjects of personal data about the transfer of their personal data to a third party. Decree of the Government of the Kyrgyz Republic. No. 759 (2017) (see https://dpa.gov.kg/en/npa/16).
About trade secrets. Law of the Kyrgyz Republic. No. 83 (1998) (see http://cbd.minjust.gov.kg/act/properties/ru-ru/38/50).
On e-governance. Law of the Kyrgyz Republic. No. 127 (2017) (see http://cbd.minjust.gov.kg/act/view/ky-kg/111634).
Cybersecurity Strategy of the Kyrgyz Republic for 2019-2023. Decree of the Government of the Kyrgyz Republic. No. 369 (2019) (see http://cbd.minjust.gov.kg/act/view/ru-ru/15478).
On Approval of the Cybersecurity Strategy of the Kyrgyz Republic for 2019-2023. On amendments to the Decree of the Government of the Kyrgyz Republic. No. 199 (2022) (see http://cbd.minjust.gov.kg/act/view/ru-ru/159115).
On Approval of the requirements for the protection of information contained in databases of state information systems. Decree of the Government of the Kyrgyz Republic. No. 762 (2017) (see https://dpa.gov.kg/en/npa/17).
On Approval of the requirements for the protection of information contained in the database of state information systems. Amendments to the Decree of the Government of the Kyrgyz Republic. No. 45 (2022) (see http://cbd.minjust.gov.kg/act/view/ru-ru/158956?cl=ru-ru).

On some issues related to state information systems. Law of the Kyrgyz Republic. No. 744 (2019) (see http://cbd.minjust.gov.kg/act/view/ru-ru/157404?cl=ru-ru).
On Approval of the Concept of open data of the Kyrgyz Republic for the period 2022-2024. Order of the Cabinet of Ministers of the Kyrgyz Republic. No. 463-r (2022) (see http://cbd.minjust.gov.kg/act/view/ru-ru/219183).

Ensuring Data Governance and Privacy in Cardiology Practice: Complying with Regulatory Frameworks and Protecting Sensitive Patient Health Data in Kyrgyzstan

Erkin Mirrakhimov, MD, Professor, Chair of Internal Medicine at the Kyrgyz State Medical Academy and head of the Atherosclerosis and Coronary Artery Disease Department of the National Center of Cardiology and Internal Disease, Bishkek, Kyrgyzstan

Kyrgyzstan is a country located in Central Asia with a population of about 7 million people; the area of the territory is 199,951 km^2. More than three-quarters of the territory of Kyrgyzstan is occupied by mountains, with more than half of its territory located at altitudes from 1,000 to 3,000 m and about a third at altitudes from 3,000 to 4,000 m (Figure 4).

As for its citizenship, about 87% of the population is under the age of 65 years old. Every fifth inhabitant of Kyrgyzstan (20%) may die young (30–69 years old) from noncommunicable diseases, such as those of the cardiovascular system, cancer, respiratory diseases, and diabetes mellitus. More than half of deaths from all causes in Kyrgyzstan are due to cardiovascular disease (CVD). In Kyrgyzstan, more than 18,000 people die from diseases of the circulatory system every year, or about 50 people a day. Given the active digitalization in the medical field, given that patient data are stored digitally on servers, and given the high prevalence of CVD in Kyrgyzstan, control, protection, security, and limitation of unauthorized access to the personal and medical data of cardiac patients is of great importance.

Legislation has been designed to provide a legal framework for handling personal data, following internationally accepted principles and standards and aligning with the Constitution and laws of Kyrgyzstan. The objective is to safeguard individuals' rights and freedoms pertaining to the collection, processing, and utilization of their personal data. The foundational legal documents, acts, and directives are based on the nation's Constitution. These include the law on personal data protection; the protocol for acquiring subjects' consent for personal data collection and processing; the procedure and format for notifying subjects about the transfer of their personal data to third parties; requirements for the security and protection of personal data throughout its processing in information systems;[12] and the procedure for registering the data repository owner, registering the data repositories, and creating listings of the data owner's personal data in the register.[13]

---

[12] Decree of the Government of Kyrgyzstan "On Approval of the Procedure for Obtaining the Consent of the Subject of Personal Data to the Collection and Processing of [Their] Personal Data, the Procedure and Form for Notifying Subjects of Personal Data about the Transfer of their Personal Data to a Third Party." No. 759 of November 21, 2017 (see https://dpa.gov.kg/en/npa/16).
[13] Law of the Kyrgyz Republic "On Personal Information" as amended by Law No. 129 of July 20, 2017, and No. 142 of November 29, 2021 (see http://cbd.minjust.gov.kg/act/view/ky-kg/202269).

FIGURE 4 Map of Kyrgyzstan. SOURCE: United States Central Intelligence Agency. 2005. *Kyrgyzstan.* https://www.loc.gov/item/2006625308 (accessed January 24, 2024).

The main data protection regulator is the State Agency for the Protection of Personal Data, established by a decree by the president of Kyrgyzstan in 2021.[14] This agency was registered on January 10, 2022, and is part of the Cabinet of Ministers of Kyrgyzstan. The decision to share personal information lies with the subjects providing it. When an individual consents to sharing their personal data, they grant permission for its collection and processing. Such consent can be given in writing, either on paper or electronically using a valid electronic signature in accordance with Kyrgyzstani legislation.

In accordance with paragraph 21 of the Law on Personal Data, the entity responsible for the personal data repository is required to use all appropriate legal, organizational, and technical measures to prevent unauthorized, illegal, or accidental access; unauthorized modification of the subject's data; access blocking; data copying; and data declassification. Noncompliance with personal data protection regulations provides legal liability. In cases where an individual's confidentiality is breached, the affected person holds the right to compensation for damages and emotional distress. Article No. 127 of October 28, 2021, of the Criminal Code of Kyrgyzstan states that violation of the private life of an individual, in particular, the illegal collection of personal data with the aim of distributing it without the subject's consent, is punishable by community service and a fine of 20,000–50,000 Kyrgyzstani Soms (equivalent to ~230–570 USD).

[14] Law of the Kyrgyz Republic "On Personal Information as amended by Law No. 129 of July 20, 2017, and No. 142 of November 29, 2021 (see http://cbd.minjust.gov.kg/act/view/ky-kg/202269).

Cardiology patients can be observed as outpatients or inpatients in medical institutions at the secondary and tertiary levels. Passport data, including last name, first name, patronymic, year of birth, social security number, and residential address, along with medical examination data and laboratory results, are entered into a secure electronic database of the corresponding medical institution. Each medical institution has its own protocol for maintaining the security of patient data with similarities between institutions. Access to the electronic database of the relevant medical institution is available only to doctors working within the institution. Unauthorized access to equipment that stores medical and patient data impossible because it is located in a special, locked office. Lack of access to this equipment prevents illegal data information acquaintance, duplication, changing information about personal data, and/or deletion from the database.

Physicians conducting outpatient or inpatient treatment do not sign special documents on the nondisclosure of patients' medical information but are guided by the principles of medical ethics, deontology, and the preservation of medical secrecy. Also, during an outpatient visit or hospitalization, patients do not provide consent for a doctor's examination, since patients came voluntarily to an outpatient appointment or inpatient treatment. Written consent is taken for instrumental invasive studies and surgical treatment. Unauthorized persons cannot access the electronic database. The right of access to the medical information of a particular patient, except for the attending physician, is provided only to the authorized bodies defined in the law On Personal Information. For example, when the patient (or their relatives in the event of the death of the patient) appeals to the Ministry of Health, law enforcement, or judicial authorities, the information is transferred to the official commissions. In addition, the medical information of patients is transferred to the Compulsory Medical Insurance Fund after each outpatient appointment and hospitalization, where patient data is entered into an electronic database, without access for unauthorized persons. Each medical institution of Kyrgyzstan submits annual statistical data on the number of treated patients, laboratory and instrumental interventions performed, and number of deaths to the E-Health Center (CEZ) of the Ministry of Health.[15] Concurrently, only digital data are transmitted to the CEZ without indication of surnames, addresses, and social security numbers. That is, no data are transmitted by which patients can be identified.

The Ministry of Health, as an authorized body of the Government of Kyrgyzstan, has developed and started to implement in pilot mode the Digital Outpatient Card of the Patient and the Digital Health Profile. The Digital Outpatient Patient Record (DAC) is an information system that stores and manages the medical data of patients, providing real-time access to medical personnel. This system improves the efficiency of providing medical care to the population by providing prompt and convenient access to information about the health of patients, as well as reducing the bureaucratic burden on medical personnel spent on searching for and collecting information. DAC includes an electronic record of the history of patients, centralized data storage, integration with laboratory and diagnostic systems, and doctor's appointment schedules. The system allows one to create reports and analysis of incidence statistics.

The Digital Human Health Profile (DPH) is an information system that aggregates and displays data related to a person collected by public health organizations. The purpose of the creation of the DPH is to improve the quality and efficiency of medical care for the population through quick and convenient access to information about the health of patients,

---

[15] Center for Electronic Healthcare. Ministry of Health Kyrgyzstan (see http://cez.med.kg/).

reducing the time to search and collect information about the patient's health. Today, with the help of a digital solution for the population, the following data are available: the health care organization providing primary health care to which the person is attached, insurance status, history of visits to health organizations providing primary health care, history of hospitalizations, history of laboratory test results, and vaccination history. In the future, integration of all other health services into the system, such as making doctor's appointments, electronic prescription documents, access to extracts from medical records, various certificate types, and others, is planned. Continuing work to integrate this information will allow the public to access patients' medical history, laboratory test results, and diagnostic tests in one place. Continuing work to integrate this information will allow the public to access patients' medical history, laboratory test results, and diagnostic tests in one place. The integration of these systems on the state portal of electronic services, Tunduk, will provide a more convenient and complete provision of the information necessary to monitor and maintain one's health.

Given the high prevalence of CVD in Kyrgyzstan, scientific research is being actively concducted in the country. When conducting scientific research in the field of cardiology in Kyrgyzstan, researchers adhere to international rules and requirements. First of all, informed written consent is obtained from the patient, having previously familiarized the patient with the planned study. The patient is provided with the purpose of the study, what data will be collected, the possible transfer of data to a third party, and the use of medical data for publication in scientific articles. At the same time, the patient's right to withdraw from the study at any time, without giving a reason, and the anonymity of the patient's data are guaranteed. The patient data included in the studies are entered into a separate secure database, access to which is available only to the researchers participating in the study. Unlike the electronic database of medical institutions, where patients are registered for outpatient or inpatient treatment, each patient participating in the study is assigned a code, by which the patient cannot be identified. Access to the patient code and identification, along with signed informed consent forms, are stored in a separate safe, the key to which is kept by the principal investigator. Researchers sign a nondisclosure document. Patient data are stored for 15–25 years, depending on the conditions of the study. The patient who plans to be included in the study independently decides whether to participate in the trial and grants the right to researchers to collect the necessary data, process and analyze them, and include the results in study reports and publication in scientific journals. In case of participation in international studies and the need to send medical information, only those medical indicators required by the study protocol are transmitted, without sending personal data by which the patient can be identified.

Digitalization is increasingly being introduced into all spheres of everyday life in Kyrgyzstan. In recent years, the country's authorities have been paying more attention to digitalization processes. The government sets the task of introducing information and communication technologies into the activities of state bodies. Thus, within the framework of the National Development Strategy of Kyrgyzstan for 2018-2040, the task of forming an open digital society is defined and organized to implement a number of activities, including:

- Provision of digital public services, including digital government, digital local government, digital parliament, and the digital justice system in all regions of the country.

- Provision of digital social services (health, education).

Since 2021, with funding from the European Union, the project "Supporting Digitalization in Kyrgyzstan" has been implemented by a consortium led by the eGA Academy of Electronic Governance (Estonia), with the participation of HAUS (Finland), CSI-Piemonte (Italy) and the Ministry of Digital Development of Kyrgyzstan. Within the framework of this project, computer literacy is being improved among residents of the country, and digital skills are being developed to use digital solutions through the development and implementation of materials for e-learning. They are increasing the availability of digital competencies and literacy, with special attention given to young people, women, people living in disadvantaged conditions, rural residents, vulnerable social groups, socially unprotected, and low-income segments of the population.

The development and implementation of e-services continue to achieve sustainable development and community empowerment. Thus, access to public services constantly expands because of their transfer to a digital platform. Many polyclinics have introduced digital records for outpatient examinations. Patients do not need to come to the clinic and wait in line for an appointment but come right at the appointment time. As part of this project, to increase the confidentiality of personal data, the Agency for the Protection of Personal Data was created, which constantly improves cybersecurity measures for the protection of personal data, risk management, and sustainability of the digitalization system. Continuous improvement of the data protection system is needed to climb the global cybersecurity rankings.

The regulatory and legal framework relating to information technology continues to improve, including issues of cybersecurity. Personal data protection and digital potential is developing. With widespread digitalization involving sectors such as the medical field, it is necessary to constantly ensure the confidentiality and protection of the population database via online platforms for raising awareness, as well as by strengthening the government's capacity in personal data protection. Furthermore, efforts are being made to increase public confidence in the security of the storage of their personal data by introducing effective data protection and confidentiality mechanisms, such as improving procedural standards and supervisory mechanisms for controlling the processing, storage and protection of personal data and ensuring confidentiality, as well as implementing a system of sanctions for violations and negligence in accordance with international human rights standards. Additionally, a key factor for a more resilient and sustainable data protection system is to improve national cybersecurity through widespread implementation. This also helps to ensure that the current government cybersecurity strategy for the coming years and security operations are focused on the main threats and risks.

Additional work is needed to create artificial intelligence capabilities and implement them in all sectors. Artificial intelligence algorithms will help in early disease diagnosis, treatment planning, and drug development. Training health care workers on digital solutions aimed at early detection and response to events in various areas of health care continues. In modern realities, digital sovereignty of Kyrgyzstan is important to remember; that is, the country's right to determine its information policy, designate its information security, and develop national software. However, we should not think of this as digital isolation. If those involved the information resources of Kyrgyzstan do not exchange data, they will not have the capabilities that the digital world represents. Further digitalization is a key factor in the growth of many sectors of the economy, the health care system, education, and many other areas of everyday human life.

Conclusion

Kyrgyzstan is actively implementing digitalization in all spheres of life, including health care. The country has created a regulatory framework on the collection of personal information, its storage, and mechanisms for protecting confidential data, and identified responsible state structures responsible for the storage and security of the database. The Ministry of Health of Kyrgyzstan is actively implementing digitalization tools in medicine and the health care field. "Digital Outpatient Card of the Patient" and "Digital Health Profile" frameworks have been created, improved, and actively integrated into the practical activities of medical institutions. Their integration will allow the population to have online access to their medical history, the results of laboratory tests, and diagnostic tests. Information about the number of treated and deceased patients, clinical diagnoses, and other medical information flows to the E-Health Center and the Compulsory Medical Insurance Fund. At the same time, patient data are protected from unauthorized access. Work continues to improve the digital literacy of medical workers and patients with more complete coverage of medical institutions by digitalization. A procedure for electronic medical document management has been developed, which excludes the maintenance of medical documents in paper form in cases when of digital solutions can replace them. This step will allow the health care system to actively implement digital solutions, reduce duplication of information, save finances, and direct them to the development of digital infrastructure health care systems. Trainings are regularly held for health care workers to improve computer and digital literacy. The Ministry of Health is constantly monitoring compliance with the security of storing confidential information and improving the methods of protecting the database.

Disclaimer: The author is solely responsible for the content of this paper, which does not necessarily represent the views of the U.S. National Academies of Sciences, Engineering, and Medicine.

Transformation of Health Information System Regulation in Kazakhstan with the Development and Widespread Use of Digital Solutions

Zhaniya Nurgaliyeva, MD, is a medical professional with more than 10 years of experience in health information systems implementation in Kazakhstan, where she has been a driving force in e-health standardization efforts in the country.

In recent years, Kazakhstan has demonstrated consistent growth in telecommunications infrastructure, the provision of digital services, and the existing human potential. Kazakhstan secured 29th place among 193 participating United Nations countries, emphasizing its well-developed information and communications technology market, with active participation from both domestic and foreign entities (UN Department of Economic and Social Affairs, 2020). Health management information systems, electronic medical records (EMR) systems, and national databases are in operation across all medical facilities. Additionally, Kazakhstan has successfully established a national telemedicine network, embraced and began developing the Internet of Things, and introduced applications for remote patient monitoring. To maintain stability and promote widespread digitalization, regulatory pressure in the form of legislation and standards plays a vital role. The process of improving legislation is time and resource intensive, necessitating a well-defined conceptual vision of development and an understanding of the necessary regulatory steps for implementation.

In the predigital era, Kazakhstan's health care sector operated under strict legislation regulating the paper-based documentation of the medical care provided to inpatients and outpatients. This encompassed patient medical records, referrals to services or hospitalization, ambulance sheets, and other primary records of medical and administrative information. The advent of information and communication technologies in health care in the 1990s initiated the generation of aggregated reports based on the main indicators of health and health care. In general, the development of information databases in various sectors of Kazakhstan developed in accordance with requirements given by the governing state body. Kazakhstan's journey toward regulated digitalization began with a state program for the formation and development of the national information infrastructure of the Republic of Kazakhstan from 2001 to 2005, alongside the first law on informatization adopted around the same time[16] ("About informatization," 2003). These foundational documents defined a regulatory framework for the information and communication technology industry and established relationships between software vendors and government agencies. Concepts such as information processes, information services, and confidential information were also introduced for the first time.

After the adoption of the strategic program for health care development in 2005, Kazakhstan embarked on the introduction of a unified health information system (HIS), the paradigm of which was to cover the processes and reporting of medical organizations in Kazakhstan with one state computer system.[17] The next few years were devoted to the development of this solution. The next step in the development of e-health in Kazakhstan was to support a large-scale reform that enabled citizens to receive medical services in any health care organization in Kazakhstan. To enable the new health care financing system, based on the "funding following the patient" principle, numerous vertical systems have been introduced since 2009, covering the process of providing care for nosology (e.g., cancer registry, tuberculosis registry, acute coronary syndrome) as have horizontal systems, responsible for individual health care functions (e.g., hospitalization, drug provision, inpatient register).

In 2012, an audit was conducted, with the involvement of experts from the Swiss Tropical and Public Health Institute, some of the results of which were reflected in the Concept for the Development of e-health in Kazakhstan in 2013–2020.[18] The main technical gap in the state computer system was that, in addition to the existing modules of the

---

[16] On the State Program for the Formation and Development of the National Information Infrastructure of the Republic of Kazakhstan. Decree of the President of the Republic of Kazakhstan No. 573 (2001) (see https://adilet.zan.kz/rus/docs/U010000573_#z0); About Informatization. Law of the Republic of Kazakhstan No. 412 (2003) (see https://adilet.zan.kz/rus/docs/Z030000412).

[17] On the State Program for the Reform and Development of Healthcare of the Republic of Kazakhstan for 2005-2010. Decree of the President of the Republic of Kazakhstan No. 1438 (2004) (see https://adilet.zan.kz/rus/docs/U040001438).

[18] Concept for the Development of Electronic Healthcare of the Republic of Kazakhstan for 2013-2020. Decree of the President of the Republic of Kazakhstan No. 464 (2013) (see https://nrchd.kz/

system based on the "thick client," web applications were additionally developed and put into operation, which led to a violation of the principle of a single database and a single data dictionary, and to an explosive increase in the need for ensuring interoperability between systems. The absence of a strategic vision and methodological and technical standards (e.g., architectural, processes, terminology) further hindered progress; the approach was eventually abandoned.

The development of health care data systems was influenced heavily by regulatory legislative acts, which focused on regulating specific functions within the health care field. For example, vertical databases were created in accordance with the standards and clinical guidelines for organizing and providing care for a particular nosology. However, as time progressed, it became evident that the primary focus was on obtaining financial statements and generating reports on the services provided by health care organizations. This emphasis on reporting improved the quality of secondary data available. Nevertheless, a significant challenge emerged as the lack of interaction between systems necessitated that doctors simultaneously work across multiple vertical and horizontal information databases. This fragmentation of patient information across different databases hindered seamless data exchange. Furthermore, the continued reliance on paper medical records for patients paralleled the advent of electronic portals and the provision of electronic financial and statistical reporting for health care services. These disconnects posed obstacles in the journey toward a fully integrated and digitalized health care ecosystem.

The idea of patient centricity gained prominence with the adoption of the first Concept for the Development of Electronic Healthcare in Kazakhstan in 2013. The reform aimed to develop an information model that empowers medical personnel to deliver safe, high-quality, and timely services.[19] Emphasizing the significance of aligning with international standards, the integration of primary health and health care data collection emerged as a natural process (European Observatory on Health Systems and Policies et al., 2019). The construction of this model remains pending, since the primary task is the formation of a single repository with clinical data of national electronic health records in accordance with the strategic document for the development of health care until 2026.[20] However, a significant outcome of the first Concept's implementation was the demonopolization of the software vendor monopoly market. Previously, the Ministry of Health solely procured health care information databases. With the adoption of the Concept, medical facilities began to acquire these systems themselves, thereby stimulating the development of the software market tailored to meet their specific needs.

Moreover, the concept highlighted the importance of establishing an e-health standardization system, encompassing the national electronic health record (EHR) information model, regulating key classification standards, as well as defining technical norms for iden-

files/ez/%D0%9A%D0%BE%D0%BD%D1%86%D0%B5%D0%BF%D1%86%D0%B8%D1%8F%20%D0%AD%D0%97.pdf).

[19] Concept for the Development of Electronic Healthcare of the Republic of Kazakhstan for 2013-2020. Decree of the President of the Republic of Kazakhstan No. 464 (2013) (see https://nrchd.kz/files/ez/%D0%9A%D0%BE%D0%BD%D1%86%D0%B5%D0%BF%D1%86%D0%B8%D1%8F%20%D0%AD%D0%97.pdf).

[20] On Approval of the Concept for the Development of Healthcare in the Republic of Kazakhstan until 2026. Decree of the Government of the Republic of Kazakhstan No. 945 (2022) (see https://adilet.zan.kz/rus/docs/P2200000945).

tification and interaction. Despite these advancements, the existing regulations for recording and maintaining clinical and administrative data, as well as secondary reporting, remained relatively unchanged, failing to consider the already advanced HIS digital infrastructure. Thus, e-health standards were developed based on the EHR model, adhering to ISO 13940. Additionally, the standards for main health processes, including electronic referrals, prescriptions, prevention strategies, patient data exchange, diagnostic test results, and electronic consultation data management, were defined. In Kazakhstan, clinical information classification standards, such as the International Classification of Diseases 9 and 10 for coding diseases, conditions, and services, as well as Anatomical Therapeutic Classification system for medicines, are well-defined and adhered to. In 2017, the implementation of health insurance reform in Kazakhstan expedited the acquisition of EMR systems by all health care providers and currently, EMR systems are operational in all health facilities across the country.

In 2020, Kazakhstan took a significant step forward with the development of the "Code on the Health of the People and the Healthcare System."[21] Prior to this law, there were difficulties in approving digitalization standards and regulating information systems and remote medical services, as the Ministry of Health, a major coordinating body governing digital health, lacked the competence to endorse these regulations. This legislation solidified the principles of digitalization, emphasizing the primacy of standards and adherence to strategic courses in the field of digital health and protection of personal data. It served to protect and enable personal medical data, remote medical services, mobile health care, medical information systems, national EHRs, and EMRs.

With the recent development of the HIS, it is important to work to revise and improve the methodology, now more than ever. As it stands now, the country continues to operate vertical and horizontal information systems, although the concept of national EHR remains unattained. Additionally, the requirements and standards for recording and collecting medical and administrative data remain the same as they were before the introduction of information systems, thereby slowing down the modern digital data collection practice and further expansion of the digital applications. For further development, it is crucial to revise and reformat the regulation related to medical and administrative records collection and health reporting and secondary use of data in health care facilities. Lastly, it is important to build requirements for target data models with a description of all relevant attributes.

Registration of events and processes within health care facilities adheres to the Ministry of Health's guidelines on the approval of forms (templates) for health care accounting.[22] An analysis of these guidelines reveals a misalignment between the prescribed paper-based format and the current level of digitalization in health care processes. Moreover, the requirements for recorded data vary across different types of services and levels of care, necessitating a more coherent approach. To address this issue, the Ministry of Health has established a working group dedicated to reviewing existing approaches in the development of standardized minimum datasets and their corresponding requirements. This strategic shift aims to introduce a targeted health care information model that is no longer based on specific accounting forms but rather centers around the essential digital data that must

---

[21] About the Health of the People and the Health Care System. Code of the Republic of Kazakhstan No. 360-VI ZRK (2020) (see https://adilet.zan.kz/rus/docs/K2000000360).

[22] On Approval of Forms of Accounting Documentation in the Field of Healthcare. Acting order Ministry of Health of Kazakhstan No. 21579 (2020) (see https://adilet.zan.kz/rus/docs/V2000021579).

be documented within medical organizations. Such an approach would facilitate the secondary use of health care data, contingent upon the standardization of medical and administrative data, their systematic comparison, and the maintenance of these within information systems.

Additionally, the current requirements for secondary (aggregated) data necessitate alignment with international standards.[23] To enhance the credibility and utility of aggregated health care data, a more comprehensive metadata framework is vital. This entails referencing the data sources and the tools employed for data collection, be it through information systems or routine collection methods. With such enhanced metadata, the computation of health indicators becomes more accurate, enabling better-informed decision-making. As part of its commitment to fostering a robust health care ecosystem, the Ministry of Health places particular emphasis on enhancing the regulatory framework governing data structure and requirements. By implementing a health care reference data model, bolstered by digital tools at the national level, the groundwork for a highly function HIS will be laid. This model-driven approach seeks to optimize data integrity, accessibility, and interoperability, ultimately contributing to the improvement of health care services and outcomes on a broader scale.

Disclaimer: The author is solely responsible for the content of this paper, which does not necessarily represent the views of the U.S. National Academies of Sciences, Engineering, and Medicine.

References

European Observatory on Health Systems and Policies, Abishev, O., and Y. Spatayev. 2019. The future development of digital health in Kazakhstan. *Eurohealth* 25(2): 24-26. Geneva: World Health Organization. https://apps.who.int/iris/handle/10665/332524 (accessed January 24, 2024).

UN (United Nations) Department of Economic and Social Affairs. 2020. *E-government survey 2020—Digital government in the Decade of Action for Sustainable Development with addendum on COVID-19 response.* New York: United Nations. https://publicadministration.un.org/egovkb/Portals/egovkb/Documents/un/2020-Survey/2020%20UN%20E-Government%20Survey%20(Full%20Report).pdf (accessed January 24, 2024).

About the System of Protection and Exchange of Epidemiological Data in Kazakhstan

Zhanna Shapieva, Elmira Utegenova, and Ulyana Kirpicheva
All authors are currently affiliated with the Scientific-Practical Center for Sanitary-Epidemiological Expertise and Monitoring (SPC SEEM), a branch of the National Center for Public Health at the Ministry of Health in Kazakhstan.
Dr. Zhanna Shapiyeva works as a Chief Specialist and Associate Professor, Ulyana Kirpicheva is an Epidemiologist, and Elmira Utegenova is Deputy Director.

---

[23] On Approval of Forms of Accounting Documentation in the Field of Healthcare. Acting order Ministry of Health of Kazakhstan No. 21579 (2020) (see https://adilet.zan.kz/rus/docs/V2000021579).

The Importance of Epidemiological Data in the Fight Against Infectious Diseases

Epidemiological data are crucial for evidence-based strategies in managing infectious and parasitic diseases. Establishing a national system for the protection and exchange of epidemiological data is vital for effectively combating diseases, promoting public health, and fostering global scientific cooperation. In Kazakhstan, this system holds strategic importance for identifying and mitigating epidemiological threats, early detecting and controlling of epidemics, conducting scientific research, and shaping public health policy. Maintaining a state information system for biological protection facilitates data exchanges and coordinated biosafety measures. Developing a national biosafety database necessitates meticulous planning, infrastructure development, and regulatory frameworks.[24]

In Kazakhstan, collaboration among various organizations, including the Ministry of Health and the National Center for Public Health, among other institutions, is pivotal. Kazakhstan has implemented measures for safeguarding personal data information (i.e., data security and confidentiality) that follows relevant laws, such as On Personal Data and Their Protection, On the Health of the People and the Healthcare System, and the Law on Biological Safety of the Republic of Kazakhstan.[25] These laws help to establish protocols and policies, data access control, information encryption, and other technical measures. For example, the law On Personal Data and Their Protection regulates the field of personal data, determines the legal framework for activities related to the collection, processing, and protection of personal data, defines violations of the law and provides requirements for electronic systems. These are carried out in accordance with the principles of 1) Respect for the constitutional rights and freedoms of individuals; 2) Legality; 3) Confidentiality of personal data with restricted access; 4) Equality of rights of subjects, owners, and operators; and 5) Ensuring the security of individuals, society, and the state. Additionally, the law On the Health of the People and the Healthcare System regulates the protection of personal medical data of individuals, whereas the law On the Biological Safety of the Republic of Kazakhstan defines the legal basis for state regulation in the field of biosafety, and aims to prevent biological threats. These laws speak to the importance of accurate, reliable, and timely data for preventing disease transmission. Specialized information systems such as these in Kazakhstan gather and store disease-related data, health indicators, and research outcomes, including morbidity cases, patient data, and medical research data.

Current Status of the Disease Surveillance System in Kazakhstan

Kazakhstan's epidemiological surveillance has been crucial for the monitoring of infectious diseases in the public health sector. Rooted in legal regulations, this intricate framework orchestrates data collection, analysis, storage, and dissemination, empowering decision-makers to design and implement public health initiatives.[26] Various entities are

---

[24] On Biological Safety of the Republic of Kazakhstan. Law of the Republic of Kazakhstan No. 122-VII LRK (2022) (see https://adilet.zan.kz/eng/docs/Z2200000122/links).

[25] On Biological Safety of the Republic of Kazakhstan. Law of the Republic of Kazakhstan No. 122-VII LRK (2022) (see https://adilet.zan.kz/eng/docs/Z2200000122/links); On the National Security of the Republic of Kazakhstan. Law of the Republic of Kazakhstan No. 527-IV (2012).

[26] About the Health of the People and the Health Care System. Code of the Republic of Kazakhstan No. 360-VI ZRK (2020) (see https://adilet.zan.kz/rus/docs/K2000000360).

involved in the surveillance system, including medical, laboratory, and sanitary-epidemiological services in both the public and private sectors. Subdivisions such as the Committee for Sanitary and Epidemiological Control (CSEC) of the Ministry of Health oversee data collection at different levels (e.g., district, city, regional), while the Scientific-Practical Center for Sanitary-Epidemiological Expertise and Monitoring (SPC SEEM) analyzes and transmits data to the CSEC and others for higher-level decision-making (SPC SEEM, 2023). Other organizations within the Ministry of Health, such as the Republic-Level AIDS Center and the National Scientific Center for Phthisiopulmonology, conduct disease surveillance. The scope of monitored health events includes communicable and noncommunicable diseases, mortality, pregnancy-related conditions and childbirth, and injuries. The country has 17 oblasts and 3 cities of republic-level significance, including Astana, Almaty, and Shymkent, that function as administrative territories for reporting purposes.

In recent years, significant resources have been invested in Kazakhstan to strengthen the surveillance system and enhance public health threat responses. Funding comes primarily from state and local budgets and is supplemented by research grants and support from international organizations, such as the U.S. Centers for Disease Control and Prevention and the Defense Threat Reduction Agency. However, there are challenges due to an outflow of qualified personnel to nonstate medical organizations and a shortage of competent personnel in state organizations, especially at the district level.

The main tasks of surveillance are:

- Timely and accurate registration of disease cases.
- Operational epidemiological analysis of population incidence.
- Detection and investigation of disease outbreaks.
- Planning, evaluating, and implementing preventive measures.
- Diagnosis, treatment, and monitoring of patients.
- Epidemiological evaluation of regions to identify high-risk populations.
- Monitoring of reservoir hosts and carriers of infectious diseases.
- Population-wide vaccination and targeted seroprophylaxis for diseases.

A tiered reporting system employs e-mail, paper, and telephone communication. Notably, Kazakhstan lacks a standardized list of case definitions for notifiable conditions, relying on clinicians' diagnoses based on disease-specific criteria. Patients meeting the criteria undergo laboratory diagnostics at designated centers, such as the National Center of Expertise. Suspected cases of infectious and parasitic diseases trigger an emergency notification to the territorial subdivision of CSEC, prompting investigations. An epidemiological survey card is completed, with findings then reported to both CSEC and SPC SEEM.[27] Kazakhstan has been embracing a digital health care system, with 31 medical and 15 laboratory information systems registered that meet quality standards such as the system of guaranteed volume of free medical care and compulsory social health insurance. These systems facilitate efficient electronic document management, data integration from various

---

[27] Rules for Registration, Keeping Records of Cases of Infectious, Parasitic, Occupational Diseases and Poisoning, and the Rules for Reporting on Them. Order of the Ministry of Health of Kazakhstan No. ҚR DSM-127 (2019); Standards in the Field of Medical Activity for Determining Cases of Especially Dangerous Human Infections During their Registration. Acting Order of the Ministry of Health of Kazakhstan No. 623 (2006).

sectors and fields, flexible patient interactions, staff oversight, and financial and organizational control. However, the absence of a unified electronic surveillance system in the CSEC system hinders the rapid acquisition of health data and identification of infectious disease clusters.

Effectiveness of the System of Protection and Exchange of Epidemiological Data in the Surveillance of Infectious Diseases

Through a well-established surveillance system, Kazakhstan can have a more proactive stance in monitoring and disease outbreak detection. The surveillance system facilitated effective strategies for case detection, treatment, and preventive measures, culminating in successful disease control. The country has encountered periodic outbreaks, such as malaria, cutaneous leishmaniasis, and measles (WHO, 2020). Notably, Kazakhstan was recognized by the World Health Organization (WHO) as a malaria-free country in 2012 (Baranova et al., 2013). In recent times, Kazakhstan, akin to other countries, has grappled with the global COVID-19 pandemic.

The surveillance system has played a key role in controlling the spread of the SARS-CoV-2 virus, and its effectiveness has been proven by several success stories. Swift responses, facilitated by real-time data exchange; included laboratory tests, contact tracing, and morbidity forecasting. Additionally, prioritizing mass vaccinations became essential. Due to detailed monitoring and data processing on registered disease cases, vulnerable population groups meriting priority access to vaccines were identified, minimizing the number of hospitalizations and fatalities (Afroogh et al., 2022). The system's success manifested in the introduction of digital platforms for contact tracing of infected individuals. While leveraging mobile applications and QR codes, it was possible to quickly identify and isolate contact persons, breaking the chain of viral transmission and reducing spread. These instances underscore the effectiveness of the surveillance system in Kazakhstan in preventing and controlling infectious disease outbreaks.

While these strides have been made, the surveillance system is not without imperfections. While the introduction of information security has expedited data transfer between medical organizations, integration with public health and the veterinary service remains incomplete. Enhanced integration would expedite data exchange within the health care system and enable prompt response to emerging public health threats. Data collection and protection in Kazakhstan are governed by regulations, ensuring the confidentiality and security of personal and epidemiological data processing. Data access is restricted to authorized persons only and encryption and tamper protection measures are in place. Yet, an absence of a unified approach and data storage system impedes the seamless integration of all the surveillance system components.

Advantages and Prospects for the Development of a Data Protection and Exchange System

The system for safeguarding and exchanging epidemiological data presents a host of advantages and prospects compared with traditional methods of data collection and processing. Real-time data transmission, for example, enables one to efficiently receive information and take timely action on epidemiological threats. Furthermore, there is also the ability to exchange data from various sources and combine them into a cohesive system. This integration results in a diminished likelihood of errors, especially those stemming from human factors, while simultaneously providing avenues for comprehensive analysis and forecasting to strengthen control and prevention measures.

Among the advantages, it is necessary to note the possibility of cross-border data exchange, bolstered by data confidentiality and security safeguards. Kazakhstan also actively collaborates with international health organizations such as WHO and CDC in the field of epidemiological data exchange. This allows Kazakhstan to access international databases, transfer its data for analysis and assessment, and participate in international research projects. Today, many of the existing systems in Kazakhstan are adequately compliant with safety requirements. However, the fact that the systems are well protected is more the merit of the developers than the standards, since there is no single vision of information security. In this regard, a single regulation on the requirements for information security within the framework of e-health of Kazakhstan will be developed. This will improve quality and lead to the unification of information security. When developing the regulations, information security specialists, suppliers, and system developers will be involved.

Obstacles and Solutions

The digitalization of health care in Kazakhstan encounters challenges such as personnel shortages, an increase in medical information, and underfunding of the health care sector. There is a lack of qualified specialists, insufficient information technology equipment and infrastructure, an excess of providers of medical information systems, and many integrations with the Ministry of Health, with many work interruptions. The lack of a single space for health care information exacerbates the situation. These problems can be solved by introducing a service model, which involves equipment and software rental ("About the Health of the People and the Health Care System," 2020; "On the National Security of the Republic of Kazakhstan," 2012; "On the Concept of Information Security of the Republic of Kazakhstan until 2016," 2011).[28] Additionally, along with the introduction of information technology in the public health service, there is still the transmission of data on cases of registration of diseases by paper (emergency notification) or by telephone. Undoubtedly, the accepted practice complicates data transfer and takes up the working time of performers for analysis and taking timely measures.

Kazakhstan has proactive and effective measures for health care digitalization, including seamless combination of e-health passport data, regardless of location. The eGov mechanism, an interface between the state and citizens, offers access to a variety of information, including vaccination records, clinical service history, electronic prescriptions, and hospitalization details. Furthermore, remote health monitoring, which was launched during the COVID-19 pandemic, is operational and developing, complemented by citizens' use of mobile applications and advancements in telemedicine. Ambulance services are also moving toward digitalization, with information going directly to e-health passports and doctors. This is augmented using artificial intelligence systems, laboratory information systems, picture archiving and communication system, and virtual reality.

[28] About the Health of the People and the Health Care System. Code of the Republic of Kazakhstan No. 360-VI ZRK (2020) (see https://adilet.zan.kz/rus/docs/K2000000360); On the National Security of the Republic of Kazakhstan. Law of the Republic of Kazakhstan No. 527-IV (2012); On the Concept of Information Security of the Republic of Kazakhstan until 2016. Decree of the President of the Republic of Kazakhstan No. 174 (2011).

In parallel, the Ministry of Health is working to improve the requirements to ensure the security of personal medical data. The regulatory legal acts preside over the responsibility of health care subjects, ensuring the confidentiality of medical personal data, delimiting access rights to medical information in systems, and revisions for the use of artificial intelligence technologies. Additionally, the Ministry of Health is working to integrate its own information systems with that of private medical entities to correctly generate accounting and reporting documentation, monitor and analyze information, and make effective management and clinical decisions. In addition, work is underway to integrate with the information systems of other government agencies for rapid data exchange. Now all systems have been transferred to the cloud, and integration is underway between them. This is not yet the integration that doctors would like, but some part of it is already in place, and work continues. As reality shows, the introduction of IS in health care has practically reduced paperwork and increased efficiency. In 2019, the introduction of health information systems in Kazakhstani health care organizations amounted to 65.1% and this figure is growing every year (Gulis et al., 2021). The main ways to solve the identified obstacles that the system of protection and exchange of epidemiological data in the country faces, in the authors' opinion, are ensuring confidentiality and cybersecurity, developing common standards, integrating existing information systems, and increasing investment in developing the effectiveness of the epidemiological information protection system.

Conclusion

The system of protection and exchange of epidemiological data in Kazakhstan is an important tool in the fight against infectious and parasitic diseases and in the protection of public health. Its effectiveness and development prospects testify to the importance of information technologies and systematic approaches in this area. Further development and improvement of the system will contribute to more effective control of epidemics and improve the quality of health care in the country. Currently, work continues on the regulation of health care digitalization, including issues of data access, storage, privacy protection, and quality assurance of technologies and software products used. Qualitative, reliable, complete, and timely digital data will become important resources of the health system.

To improve and develop the system of protection and exchange of epidemiological data in Kazakhstan, it is necessary to increase investments in the digitalization of health care, strengthen human resources, improve the regulatory framework, and develop partnerships with international organizations. This will help improve data quality and management, surveillance and laboratory information systems, response to epidemic threats, disease forecasting; it will also promote research and development of effective public health interventions. Continuous improvement of the data protection and exchange system and its integration with the international scientific community will contribute to more effective epidemic control, disease prevention, and public health protection in Kazakhstan.

Disclaimer: The author is solely responsible for the content of this paper, which does not necessarily represent the views of the U.S. National Academies of Sciences, Engineering, and Medicine.

References

Afroogh, S., A Esmalian, A. Mostafavi, A. Akbari, K. Rasoulkhani, S. Esmaeili, and E. Hajiramezanali. 2002. Tracing app technology: An ethical review in the COVID-19 era and directions for post-COVID-19. *Ethics and Information Technology* 24(3): 30. https://doi.org/10.1007/s10676-022-09659-6.

Baranova, A. M., M. N. Ezhov, T. M. Guzeeva, and L. F. Morozova LF. 2013. [The current malaria situation in the CIS countries (2011-2012)]. *Med Parazitol (Mosk)* 4:7-10. Russian. PMID: 24640123.

Gulis, G., A. Aringazina, Z. Sangilbayeva, K. Zhan, E. de Leeuw, and J. P. Allegrante. 2021. Population health status of the republic of Kazakhstan: Trends and implications for public health policy. *International Journal of Environmental Research and Public Health* 18(22):12235.

SPC SEEM (Scientific and Practical Center of Sanitary-Epidemiological Examination and Monitoring). n.d. *Ministry of Health*. https://rk-ncph.kz.

WHO (World Health Organization). 2020. *Cross-border collaboration on malaria between countries of the WHO Eastern Mediterranean and European regions: Report of the 98ioregional coordination meeting Dushanbe, Tajikistan.* WHO/EURO No. 2020-1030-40776-55007. Geneva: World Health Organization.

Data Management in Life Sciences in Kazakhstan: Balancing Privacy, Security, and the Development of Open Science

Pavel Tarlykov, Head of Proteomics and Mass Spectrometry Laboratory,
National Center for Biotechnology (Astana, Kazakhstan)

Introduction and Overview of the Problem

Throughout history, human curiosity has driven the systematic collection of information about the world around us. From ancient rock paintings to modern writings, the accumulation of knowledge in the form of clay tablets, papyri, and finally books has led to the establishment of libraries and archives (Schmandt-Besserat, 1979). Moreover, access to certain information storage centers had its limitations; for example, those who could not read, or did not have the proper qualifications, could not access these spaces. However, millennia later, the problem of using and disseminating information has become much more complicated, facilitated by the advent of technology and groundbreaking scientific discoveries.

Scientific progress led to the emergence of computers, the discovery of the DNA structure (Watson and Crick, 1953), and the acceptance of the central dogma of molecular biology (Crick, 1970). Over the past 40 years, new scientific fields such as genomics and proteomics emerged, now better known as -omics technologies (Yadav, 2007), generating vast amounts of data. The classification of data is based on several parameter types, including biomolecule type (e.g., DNA, RNA, proteins, metabolites), organism (e.g., bacteria, archaea, eukaryotes), pathogenicity, and application (e.g., clinical data, environmental sample data), among others. Some data types have more restrictions on their use and dissemination than others, as the exposure of some represents potential risks, while other data types may be personal, with prohibited distribution unless proper consent from the data subject is obtained. This has highlighted the need for different levels of privacy, security, and openness in data management, depending on the data type and its potential implications.

Although the data types obtained are similar across countries, the mechanisms governing their use and dissemination oftentimes vary. The specific regulations for processing and handling data depends largely on the state's level of development. Typically, the more developed the state, the better regulated the data management processes. Central Asian countries, including Kazakhstan, are currently still in the process of development, having gained independence relatively recently. This has resulted in gaps in data management infrastructure, hindering data exchange, standardization, and quality control. Another drawback is the issue of compliance with existing legislations. This often occurs when there are ambiguities in the law or provisions that can be interpreted in different ways. It takes time and resources to tune these processes, especially in the case of low to moderate economic income in Central Asian countries. However, there are existing legislative frameworks that may serve as the basis for the development of subsequent, more specific, regulations in these countries.

Balancing Privacy, Security, and Open Science Advancements

Data Privacy

Since 2013, Kazakhstan has enacted the law On Personal Data and Their Protection, which defines personal and biometric data and outlines their use, storage, and dissemination principles.[29] Article 7 of the law specifically addresses the conditions for the collection and processing of personal data, allowing data dissemination in public sources with the consent of the data subject or their legal representative. This point safeguards personal information and establishes guidelines for processing personal data in scientific research.

Furthermore, the law On Public Health and Healthcare Systems details personal medical data. This legislation regulates the collection and management of health-related data, ensuring their confidentiality and proper use through procedures such as informed consent, which is a procedure for a person to voluntarily confirm their consent to receive medical care or participate in a research study. The adoption of this code, regulating ethical and legal conduct within medical and biological research, was an important step for the biomedical science field in Kazakhstan as it directly protected patients' rights. This existing law established a foundation for data management regulation, primarily safeguarding the confidentiality of human and patient data. Another relevant law, the Law on Access to Information, adopted in 2015, addresses information with restricted access, including state secrets, personal, and legally protected data. This law enhances data transparency and accessibility in biomedical research, striking a balance between information availability and privacy protection by restricting access to information that could cause harm if exposed, such as information related to national security and defense and public health, among other interests.

---

[29] On Personal Data and Their Protection. The Law of the Republic of Kazakhstan No. 94-V (2013) (see https://adilet.zan.kz/eng/docs/Z1300000094#:~:text=The%20Law%20of%20the%20Republic,94%2DV.&text=This%20Law%20regulates%20the%20public,and%20protection%20of%20personal%20data).

Data Security

Given the diverse range of research subjects, including potential pathogenic biological agents, along with the increasing threats to biosecurity, such as the COVID-19 pandemic caused by the spread of the SARS-CoV-2 virus, Kazakhstan recognized the need for increased biosecurity regulations. In 2022, the law On Biological Safety of the Republic of Kazakhstan was adopted to establish heightened control for work related to pathogenic biological agents for nonmilitary purposes.[30] The law requires researchers to comply with legal requirements when working with restricted information related to pathogens. Additionally, it establishes the roles and responsibilities of certain governing bodies, such as the Ministry of Defense of Kazakhstan's responsibility for interdepartmental coordination of measures to ensure biological safety and the Ministry of Health's responsibility for the health, sanitary, and epidemiological welfare of the population. Additionally, the law promotes international data exchange, in accordance with international treaties, to prevent biological threats. This point is especially important considering the COVID-19 pandemic, where timely and consistent exchange of biomedical information at the international level was essential.

In the cases where a scientific researcher works with nonpathogenic biological organisms not regulated by these national laws, the use and dissemination of the data obtained should be regulated by the internal organization in which the study is being conducted. The regulations pertaining to data classification, in part, come from the state funding body (e.g., the Ministry of Defense or the Ministry of Health of Kazakhstan). Oftentimes, one of the requirements of the funding body is the publication of results in peer-reviewed journals indexed by international databases (e.g., Web of Science and Scopus). This implies data use and dissemination in biomedical research with living organisms. While publishing in peer-reviewed journals is considered generally accessible to some in Kazakhstan, researchers with concerns about the publication of sensitive information should seek approval from the ethics committee, the academic council of the research institute or university, and/or other governing body.

Data in the Age of Open Science

In line with presenting the principles of open science, we present two research studies on proteomics and mass spectrometry conducted by the National Center for Biotechnology (NCB) in Astana, Kazakhstan. These studies offer insights into the prevailing directions of the medical and biological research fields in the Central Asian region, particularly emphasizing the paramount considerations of data confidentiality and security.

The first study encompasses whole-genome sequencing of clinical isolates of *Mycobacterium tuberculosis* in Kazakhstan (Tarlykov et al., 2020). The research centered on M. tuberculosis DNA isolates obtained from the City Center for Phthisiopulmonology in Astana, Kazakhstan. To obtain the biomaterial, a memorandum of cooperation between the City Center for Phthisiopulmonology and the NCB was signed, and permission was obtained from the local ethical commission to conduct the study. Measures were undertaken to depersonalize patient data, withholding any personally identifiable information from the researchers. An interesting feature of the *M. tuberculosis* isolates under study was their

---

[30] On Biological Safety of the Republic of Kazakhstan. Law of the Republic of Kazakhstan No. 122-VII LRK (2022) (see https://adilet.zan.kz/eng/docs/Z2200000122/links).

belonging to a subfamily of the L4 genetic line called LAM-RUS, which is endemic for the Commonwealth of Independent States countries. Understanding the genomes from this subgroup was important from an epidemiological standpoint, as it relates to the rapid acquisition of antibiotic resistance among LAM-RUS isolates. The completed genome sequences were uploaded to GenBank databases (https://www.ncbi.nlm.nih.gov/genbank/) and Sequence Read Archive (https://www.ncbi.nlm.nih.gov/sra) (Tarlykov et al., 2020).

The second study delved into the analysis of interactions among pluripotent transcription factors, including the SRY-Box Transcription Factor 2 (Sox2), octamer-binding protein-4 (Oct4), and Homeobox protein (Nanog), utilizing mass spectrometry–based proteomic data (Kulyyassov and Ogryzko, 2020). A quantitative analysis of the interactions of the Sox2, Oct4, and Nanog transcription factors was carried out through in vivo mass spectrometry. The raw data were uploaded to the Proteomics Identifications Database: PRIDE repository (https://www.ebi.ac.uk/pride/). This aligned with the requirement to publish data in an open repository for consideration in peer-reviewed scientific journals (Kulyyassov and Ogryzko, 2020). The study focused on the commercial cell line HEK293T, obtained from human embryonic kidney cells in the early 1970s, warranting no ethical approval or permits for the investigation. Mass spectrometric analysis was conducted at the NCB. Notably, the experiments offered a novel approach based on a biotinylation technique, involving a target-enzyme, BAP/BirA, system to label target proteins in their proximal interactions, presenting a valuable alternative to conventional methods for analyzing protein interaction. Furthermore, additional biomedical information, such as ChIP-Seq data and human STR data, has been uploaded into open repositories at the NCB (Jurisic et al., 2018; Zhabagin et al., 2020).

Kazakhstan As It Relates to International Data Governance

Considering current trends toward both political and economic globalization, fostering international cooperation for data use and dissemination, underpinned by international treaties, emerges as the favorable path for Kazakhstan. The Cartagena Protocol serves as a prominent example of successful cooperation. Kazakhstan's engagement with the Cartagena Protocol began with its accession to the Convention on Biological Diversity in 1994, followed by the ratification of the Cartagena Protocol in 2008. Subsequently, in 2008, the government of Kazakhstan appointed the Ministry of Agriculture at the coordinating center, and the Ministry of Education and Science as the competent national body.[31]

Moreover, in 2009, by order of the Ministry of Education and Science, the NCB was designated to serve as a focal point for the Biosafety Clearing-House, a mechanism under the Cartagena Protocol on Biosafety.[32] This platform facilitates the exchange of scientific, technical, environmental, and legal information related to living modified organisms (LMOs) and genetically modified objects (GMOs). The NCB also provides valuable knowledge on the Cartagena Protocol on Biosafety to the Convention on Biological Diversity, including regulations, strategic plans for the protocol, guidelines for risk assessment of LMOs, and an information global database on GMOs.

---

[31] On Measures to Ensure the Fulfillment by the Republic of Kazakhstan of Obligations Arising from the Cartagena Protocol on Biosafety to the Convention on Biological Diversity. Decree of the Government of the Republic of Kazakhstan No. 1282 (2008) (see https://bch.cbd.int/en/database/102592).

[32] Order of the Ministry of Education and Science of the Republic of Kazakhstan No. 579 (2009).

Drawing from this experience, the establishment of a collaborative working group among key stakeholders, such as the Ministry of Health and the Ministry of Science and Higher Education, the National Academy of Sciences of Kazakhstan, the Pugwash Committee, and other interested organizations, represents the initial stride toward engaging with the international community on data use, privacy, security concerns, and dissemination. It is of the utmost importance for scientists involved in medical and biological research in Kazakhstan to vigilantly stay abreast of legislative changes. Additionally, adhering to international ethical standards and best practices in life sciences data management is fundamental. By embracing international cooperation and aligning with established national and international protocols, Kazakhstan can foster an environment that ensures responsible and transparent use and dissemination of data.

Disclaimer: The author is solely responsible for the content of this paper, which does not necessarily represent the views of the U.S. National Academies of Sciences, Engineering, and Medicine.

References

Crick, F. 1970. Central dogma of molecular biology. *Nature* 227(5258):561-563. https://doi.org/10.1038/227561a0.

Jurisic, A., C. Robin, P. Tarlykov, L. Siggens, B. Schoell, A. Jauch, . . . and V. Ogryzko. 2018. Topokaryotyping demonstrates single cell variability and stress dependent variations in nuclear envelop associated domains. *Nucleic Acids Research* 46(22): e135. https://doi.org/10.1093/nar/gky818.

Kulyyassov, A., and V. Ogryzko. 2020. In vivo quantitative estimation of DNA-dependent interaction of sox2 and oct4 using bira-catalyzed site-specific biotinylation. *Biomolecules* 10(1). https://doi.org/10.3390/biom10010142.

Schmandt-Besserat, D. 1979. An archaic recording system in the Uruk-Jemdet Nasr Period. *American Journal of Archaeology* 83(1):19-48. https://doi.org/10.2307/504234.

Tarlykov, P., S. Atavliyeva, A. Alenova, and Y. Ramankulov. 2020. Genomic analysis of Latin American-Mediterranean family of mycobacterium tuberculosis clinical strains from Kazakhstan. *Memorias do Instituto Oswaldo Cruz* 115(8):1-6. https://doi.org/10.1590/0074-02760200215.

Watson, J. D., and F. H. Crick. 1953. Molecular structure of nucleic acids; A structure for deoxyribose nucleic acid. *Nature* 1719(4356):737-738. https://doi.org/10.1038/171737a0.

Yadav, S. P. 2007. The wholeness in suffix –omics, -omes, and the word om. *Journal of Biomolecular Techniques* 18(5):277. https://doi.org/10.3390/ijms20194781.

Zhabagin, M., Z. Sabitov, P. Tarlykov, I. Tazhigulova, Z. Junissova, D. Yerezhepov, . . . and E. Balanovska. 2020. The medieval Mongolian roots of Y-chromosomal lineages from South Kazakhstan. *BMC Genetics* 21(87):87. https://doi.org/10.1186/s12863-020-00897-5.

Appendix F

Data Governance Resources

The following resources on data governance were shared by committee members, panelists, or other participants during the workshop series.

International Resources

- The Recommendation on Science and Scientific Researchers (2017)[1] and Recommendation on Open Science (2021)[2] of the United Nations Educational, Scientific and Cultural Organization (UNESCO), a human rights-based organization that promotes open science and benefits sharing.
- The Sharing and Reuse of Health-related Data for Research Purposes: WHO Policy and Implementation Guidance (2022)[3] of the World Health Organization (WHO), an agency of the United Nations that connects nations, partners, and people to promote health, keep the world safe, and serve the vulnerable. WHO published guidance for sharing and reuse of health data for research purposes based on lessons learned from the pandemics and is currently working on an international pandemic treaty to facilitate data and sample sharing during times of pandemic.
- The Nagoya Protocol on Access to Genetic Resources and the Fair and Equitable Sharing of Benefits Arising from their Utilization,[4] a supplementary agreement to the Convention on Biological Diversity. The convention and protocol only apply to nonhuman biological samples and genetic resources and establish a specific, procedural system of benefits sharing when sharing genetic resources.
- The FAIR Guiding Principles for Scientific Data Management and Stewardship (2016)[5], a set of technical principles to improve the Findability, Accessibility, Interoperability, and Reusability of digital assets. These principles describe aspirational technical requirements needed to share data at the international level.
- The Beacon API,[6] Framework for Responsible Sharing of Genomic and Health-Related Data,[7] and Data Use Ontology[8] of the Global Alliance for Genomics & Health (GA4GH), an organization that promotes and facilitates international sharing of genomic and health data.

---

[1] See https://en.unesco.org/themes/ethics-science-and-technology/recommendation_science.
[2] See https://en.unesco.org/science-sustainable-future/open-science/recommendation.
[3] See https://www.who.int/publications/i/item/9789240044968.
[4] See https://www.cbd.int/abs/about/.
[5] See https://www.go-fair.org/fair-principles/.
[6] See https://beacon-project.io/.
[7] See https://www.ncbi.nlm.nih.gov/pmc/articles/PMC4685158/.
[8] See https://www.ga4gh.org/our-products/.

- The framework for the promotion of ethical benefit sharing in health research[9] (2022), PHA4GE hAMRonization Tool,[10] and COVID-19 Metadata Template[11] of the Public Health Alliance for Genomic Epidemiology (PHA4GE), the nonhuman equivalent organization to GA4GH, for pathogens and microbes.

National and Regional Resources

- In the European Union, the General Data Protection Regulation[12] establishes rules for the processing and transfer of personal data.
- In Kazakhstan, the Law on Informatization,[13] the Law on Personal Data and Their Protection,[14] and the Code on the Health of the People and the Healthcare System[15] provide regulatory frameworks for the information, communication technology, and medical industry. Also, the Law on Biological Safety of the Republic of Kazakhstan[16] requires compliance with legal requirements when working with restricted pathogen information.
- In Kyrgyzstan, the National Statistical Committee communication strategy helps ensure data access and security; a database being launched by the Ministry of Digital Development aims to improve data management and information protection. The Law on Electronic Governance defines the rights and obligations of data owners, and the Law on the Conflict of Interest addresses national security, transparency, and control, as well as personal responsibility and liability.
- In India, the Digital Personal Data Protection Bill[17] is a proposed law to address the management of digitized personal data.
- In Taiwan, the National Health Insurance Research Database[18] links more than 400 sources of data and establishes data protections and access protocols to protect security and confidentiality.
- In Tajikistan, the Law on Protection of Personal Data; Law on Ensuring Biological Safety, Biological Security, and Biological Protection; Law on Genetic Resources; and Healthcare Code establish frameworks for data collection, management, and use of various types of biological data.
- In Uganda, the National Biorepository[19] establishes a workflow and requirements for storing and accessing biological samples and data.

---

[9] See https://gh.bmj.com/content/7/2/e008096.
[10] See https://github.com/pha4ge/hAMRonization.
[11] See https://docs.google.com/spreadsheets/d/17PuBcA0cCT-j9hV5tbwMFKtwWwKE-a_MYRqOOsIxj7c/edit#gid=136997361.
[12] See https://gdpr.eu/.
[13] See https://adilet.zan.kz/rus/docs/Z030000412_.
[14] See https://adilet.zan.kz/eng/docs/Z1300000094#:~:text=The%20Law%20of%20the%20Republic,94%2DV.&text=This%20Law%20regulates%20the%20public,and%20protection%20of%20personal%20data.
[15] See .https://adilet.zan.kz/rus/docs/K2000000360.
[16] See https://adilet.zan.kz/eng/docs/Z2200000122/links.
[17] See https://prsindia.org/billtrack/draft-the-digital-personal-data-protection-bill-2022.
[18] See https://nhird.nhri.edu.tw/en/.
[19] See https://www.ncbi.nlm.nih.gov/pmc/articles/PMC7479498/.

- In the United States, the National Institutes of Health scientific data sharing resources,[20] Genomic Data Sharing Policy,[21] and Data Management and Sharing Policy[22] provide mechanisms for management and sharing of data from federally funded research.
- In Uzbekistan, the National Council on Biological Safety is charged with protecting life science data throughout its life cycle.

[20] See https://sharing.nih.gov/.
[21] See https://sharing.nih.gov/genomic-data-sharing-policy.
[22] See https://sharing.nih.gov/data-management-and-sharing-policy.